McGraw-Hill

Dictionary of
Astronomy

Second
Edition

McGraw-Hill

New York   Chicago   San Francisco   Lisbon   London   Madrid
Mexico City   Milan   New Delhi   San Juan   Seoul   Singapore
Sydney   Toronto

All text in the dictionary was published previously in the McGRAW-HILL DICTIONARY OF SCIENTIFIC AND TECHNICAL TERMS, Sixth Edition, copyright © 2003 by The McGraw-Hill Companies, Inc. All rights reserved.

McGRAW-HILL DICTIONARY OF ASTRONOMY, Second Edition, copyright © 2003 by The McGraw-Hill Companies, Inc. All rights reserved. Printed in the United States of America. Except as permitted under the United States Copyright Act of 1976, no part of this publication may be reproduced or distributed in any form or by any means, or stored in a database or retrieval system, without the prior written permission of the publisher.

1 2 3 4 5 6 7 8 9 0   DOC/DOC   0 9 8 7 6 5 4 3

ISBN 0-07-141047-3

This book is printed on recycled, acid-free paper containing a minimum of 50% recycled, de-inked fiber.

This book was set in Helvetica Bold and Novarese Book by the Clarinda Company, Clarinda, Iowa. It was printed and bound by RR Donnelley, The Lakeside Press.

McGraw-Hill books are available at special quantity discounts to use as premiums and sales promotions, or for use in corporate training programs. For more information, please write to the Director of Special Sales, Professional Publishing, McGraw-Hill, Two Penn Plaza, New York, NY 10121-2298. Or contact your local bookstore.

Library of Congress Cataloging-in-Publication Data

McGraw-Hill dictionary of astronomy — 2nd. ed.
   p. cm.
  "All text in this dictionary was published previously in the McGraw-Hill dictionary of scientific and technical terms, sixth edition, c2003..."
— T.p. verso.
  ISBN 0-07-141047-3 (alk. paper)
  1. Astronomy—Dictionaries. 2. Astrophysics—Dictionaries.
3. Aerospace engineering—Dictionaries. I. Title: Dictionary of astronomy.
II. McGraw-Hill dictionary of scientific and technical terms. 6th ed.

QB14.M36 2003
520'.3—dc21

                    2002033198

Contents

Preface

The *McGraw-Hill Dictionary of Astronomy* provides a compendium of more than 2800 terms that are central to the discipline of astronomy and related fields of science and technology. The coverage in this Second Edition is focused on the core fields of astronomy, astrophysics, and aerospace engineering related to space exploration. New terms have been added and others revised as necessary, and less relevant terminology in the areas of military and civilian aerospace engineering have been transferred to companion dictionaries.

Astronomy may be defined as the scientific study of the universe and the objects in it. It draws heavily on physics, and the terms "astronomy" and "astrophysics" are often used interchangeably. However, astronomy also utilizes the principles, methods, and technologies of many other disciplines, such as chemistry, geology, and biology. For example, spectroscopy has been a key technique for more than a century in determining the composition and even the motion of distant celestial objects. Highly sophisticated technology is employed on earth as well as in spacecraft to observe, measure, and analyze radiation received from elsewhere in the universe. The terms included in this Dictionary are fundamental to an understanding of these fields and the scientific approach of astronomy.

All of the definitions are drawn from the *McGraw-Hill Dictionary of Scientific and Technical Terms*, Sixth Edition (2003). The pronunciation of each term is provided along with synonyms, acronyms, and abbreviations where appropriate. A guide to the use of the Dictionary appears on pages vii and viii, explaining the alphabetical organization of terms, the format of the book, cross referencing, and how synonyms, variant spellings, abbreviations, and similar information are handled. The Pronunciation Key on page x will also guide the reader. The Appendix provides useful conversion tables for commonly used scientific units as well as other listings of astronomical data.

It is the editors' hope that the Second Edition of the *McGraw-Hill Dictionary of Astronomy* will serve the needs of scientists, students, teachers, librarians, and writers for high-quality information, and that it will contribute to scientific literacy and communication.

Mark D. Licker
Publisher

Staff

Mark D. Licker, Publisher—Science

Elizabeth Geller, Managing Editor
Jonathan Weil, Senior Staff Editor
David Blumel, Staff Editor
Alyssa Rappaport, Staff Editor
Charles Wagner, Digital Content Manager
Renee Taylor, Editorial Assistant

Roger Kasunic, Vice President—Editing, Design, and Production

Joe Faulk, Editing Manager
Frank Kotowski, Jr., Senior Editing Supervisor

Ron Lane, Art Director

Thomas G. Kowalczyk, Production Manager
Pamela A. Pelton, Senior Production Supervisor

Henry F. Beechhold, Pronunciation Editor
Professor Emeritus of English
Former Chairman, Linguistics Program
The College of New Jersey
Trenton, New Jersey

How to Use the Dictionary

ALPHABETIZATION. The terms in the McGraw-Hill Dictionary of Astronomy, Second Edition, are alphabetized on a letter-by-letter basis; word spacing, hyphen, comma, solidus, and apostrophe in a term are ignored in the sequencing. For example, an ordering of terms would be:

Col	**CrA**
cold dark matter	**Crab Nebula**
cold-flow test	**Crab pulsar**

FORMAT. The basic format for a defining entry provides the term in boldface, the field is small capitals, and the single definition in lightface:

> **term** |FIELD|　Definition.

A term may be followed by multiple definitions, each introduced by a boldface number:

> **term** |FIELD|　**1.** Definition. **2.** Definition. **3.** Definition.

A simple cross-reference entry appears as:

> **term**　*See* another term.

A cross reference may also appear in combination with definitions:

> **term** |FIELD|　**1.** Definition. **2.** *See* another term.

CROSS REFERENCING. A cross-reference entry directs the user to the defining entry. For example, the user looking up "Acrux" finds:

> **Acrux**　*See* α Crucis.

The user then turns to the "C" terms for the definition. Cross references are also made from variant spellings, acronyms, abbreviations, and symbols.

> **A/E ratio**　*See* absorptivity-emissivity ratio.
> **aestival**　*See* estival.
> **AGB**　*See* asymptotic giant branch.

ALSO KNOWN AS . . . , etc. A definition may conclude with a mention of a synonym of the term, a variant spelling, an abbreviation for the term, or other such information, introduced by "Also known as . . . ," "Also spelled . . . ," "Abbreviated . . . ," "Symbolized . . . ," "Derived from . . . ." When a term has more than one definition, the positioning of any of these phrases conveys the extent of applicability. For example:

term |FIELD| **1.** Definition. Also known as synonym. **2.** Definition. Symbolized T.

In the above arrangement, "Also known as . . ." applies only to the first definition; "Symbolized . . ." applies only to the second definition.

term |FIELD| Also known as synonym. **1.** Definition. **2.** Definition.

In the above arrangement, "Also known as . . ." applies to both definitions.

Fields and Their Scope

[AERO ENG] **aerospace engineering**—The branch of engineering pertaining to the design and construction of aircraft and space vehicles and of power units, and dealing with the special problems of flight in both the earth's atmosphere and space, such as in the flight of air vehicles and the launching, guidance, and control of missiles, earth satellites, and space vehicles and probes. This dictionary focuses on terms related to space exploration.

[ASTRON] **astronomy**—The science concerned with celestial bodies and with the observation and interpretation of radiation received from the component parts of the universe.

[ASTROPHYS] **astrophysics**—The branch of astronomy that treats of the physical properties of celestial bodies, such as luminosity, size, mass, density, temperature, and chemical composition, and their origin and evolution.

Pronunciation Key

Vowels
a	as in bat, that
ā	as in bait, crate
ä	as in bother, father
e	as in bet, net
ē	as in beet, treat
i	as in bit, skit
ī	as in bite, light
ō	as in boat, note
ȯ	as in bought, taut
u̇	as in book, pull
ü	as in boot, pool
ə	as in but, sofa
au̇	as in crowd, power
ȯi	as in boil, spoil
yə	as in formula, spectacular
yü	as in fuel, mule

Semivowels/Semiconsonants
w	as in wind, twin
y	as in yet, onion

Stress (Accent)
ˈ	precedes syllable with primary stress
ˌ	precedes syllable with secondary stress
⁞	precedes syllable with variable or indeterminate primary/secondary stress

Consonants
b	as in bib, dribble
ch	as in charge, stretch
d	as in dog, bad
f	as in fix, safe
g	as in good, signal
h	as in hand, behind
j	as in joint, digit
k	as in cast, brick
ḵ	as in Bach (used rarely)
l	as in loud, bell
m	as in mild, summer
n	as in new, dent
n̲	indicates nasalization of preceding vowel
ŋ	as in ring, single
p	as in pier, slip
r	as in red, scar
s	as in sign, post
sh	as in sugar, shoe
t	as in timid, cat
th	as in thin, breath
th̲	as in then, breathe
v	as in veil, weave
z	as in zoo, cruise
zh	as in beige, treasure

Syllabication
·	Indicates syllable boundary when following syllable is unstressed

A

Abell richness classes [ASTRON] A scale of six categories of richness into which clusters of galaxies are classified, based on the number of galaxies observed in the cluster that are not more than 2 magnitudes fainter than the third-brightest member. { 'ā·bəl 'rich·nəs ˌklas·əz }

aberration [ASTRON] The apparent angular displacement of the position of a celestial body in the direction of motion of the observer, caused by the combination of the velocity of the observer and the velocity of light. { ˌab·ə'rā·shən }

ablation [AERO ENG] The intentional removal of material from a nose cone or spacecraft during high-speed movement through a planetary atmosphere to provide thermal protection to the underlying structure. { ə'blā·shən }

ablative cooling [AERO ENG] The carrying away of heat, generated by aerodynamic heating, from a vital part by arranging for its absorption by a nonvital part. { 'a·blə·div 'kül·iŋ }

ablative shielding [AERO ENG] A covering of material designed to reduce heat transfer to the internal structure through sublimation and loss of mass. { 'a·blə·div 'shēld·iŋ }

abort [AERO ENG] **1.** To cut short or break off an action, operation, or procedure with an aircraft, space vehicle, or the like, especially because of equipment failure. **2.** An aircraft, space vehicle, or the like which aborts. **3.** An act or instance of aborting. { ə'bȯrt }

absolute magnitude [ASTROPHYS] **1.** A measure of the brightness of a star equal to the magnitude the star would have at a distance of 10 parsecs from the observer. **2.** The stellar magnitude any meteor would have if placed in the observer's zenith at a height of 100 kilometers. { 'ab·səˌlüt 'mag·nə·tüd }

absorption nebula *See* dark nebula. { əb'sȯrp·shən 'neb·yə·lə }

absorptivity-emissivity ratio [ASTROPHYS] In space applications, the ratio of absorptivity for solar radiation of a material to its infrared emissivity. Also known as A/E ratio. { əbˌsȯrp'tiv·əd·ē ˌē·mə'siv·ə·tē ˌrā·shō }

acceleration mechanisms [ASTROPHYS] The ways in which cosmic-ray and solar-flare particles may have acquired their high energies. { akˌsel·ə'rā·shən 'mek·ə·niz·əmz }

accretion [ASTRON] A process in which a star gathers molecules of interstellar gas to itself by gravitational attraction. { ə'krē·shən }

accretion disk [ASTRON] A viscous structure consisting of gas lost by a red giant or supergiant flowing around a companion main-sequence star or compact object (white dwarf, neutron star, or black hole). { ə'krē·shən ˌdisk }

accretion hypothesis [ASTRON] Any hypothesis which assumes that the earth originated by the gradual addition of solid bodies, such as meteorites, that were formerly revolving about the sun but were drawn by gravitation to the earth. { ə'krē·shən hī'päth·ə·səs }

accretion theory [ASTRON] A theory that the solar system originated from vortices in a disk-shaped mass. { ə'krē·shən 'thē·ə·rē }

Achilles [ASTRON] An asteroid; member of the group known as the Trojan planets. { ə'kil·ēz }

Achilles group *See* Greek group. { ə'kil·ēz ˌgrüp }

Acrux *See* α Crucis. { 'āˌcrəks }

actinometry

actinometry [ASTROPHYS] The science of measurement of radiant energy, particularly that of the sun, in its thermal, chemical, and luminous aspects. { ‚ak·tə'näm·ə·trē }

active center [ASTRON] A localized, transient region of the solar atmosphere in which sunspots, faculae, plages, prominences, solar flares, and so forth are observed. { 'ak·tiv 'sen·tər }

active galactic nucleus [ASTRON] A central region of a galaxy, a light-year or less in diameter, where violent and apparently explosive behavior is observed which is manifested in many ways, including the high-velocity outflow of gas, strong nonthermal radio emission, intense and often polarized and highly variable radiation over a wide range of wavelength bands, and ejection of jets of relativistic material. { ¦ak· tiv gə‚lak·tik 'nü·klē·əs }

active galaxy [ASTRON] A galaxy whose central region exhibits strong emission activity, from radio to x-ray frequencies, probably as a result of gravitational collapse; this category includes M82 galaxies, Seyfert galaxies, N galaxies, and possibly quasars. { 'ak·tiv 'gal·ək·sē }

active prominence [ASTRON] A classification of prominences of the sun; such a prominence is rapidly moving, and is the most frequent type. { 'ak·tiv 'präm·ə·nəns }

active prominence region [ASTRON] Portions of the solar limb that display active prominences, characterized by down-flowing knots and streamers, sprays, frequent surges, and curved loops. Abbreviated APR. { 'ak·tiv 'präm·ə·nəns ‚rē·jən }

active region [ASTRON] A localized, transient, nonuniform region on the sun's surface, penetrating well down into the lower chromosphere. { 'ak·tiv 'rē·jən }

active satellite [AERO ENG] A satellite which transmits a signal. { 'ak·tiv 'sad·ə‚līt }

active Sun [ASTRON] The Sun during the portion of its 11-year cycle in which sunspots, flares, prominences, and variations in radio-frequency emission reach their maximum. { 'ak·tiv 'sən }

actual exhaust velocity [AERO ENG] **1.** The real velocity of the exhaust gas leaving a nozzle as determined by accurately measuring at a specified point in the nozzle exit plane. **2.** The velocity obtained when the kinetic energy of the gas flow produces actual thrust. { 'ak·chə·wəl ig'zòst və‚läs·əd·ē }

adapter skirt [AERO ENG] A flange or extension of a space vehicle that provides a ready means for fitting some object to a stage or section. { ə'dap·tər ‚skərt }

Adhara [ASTRON] A star of spectral type B2II. Also known as ε Canis Majoris. { ə'där·ə }

adiabatic approximation [ASTROPHYS] The approximation that the pressure and density of gas in a star are related by the adiabatic law. { ¦ad·ē·ə¦bad·ik ə‚präk·sə'mā·shən }

Adonis [ASTRON] An asteroid with an orbital eccentricity of 0.779 and a perihelion well inside the orbit of Venus that passed about 1×10^6 miles (1.6 × 10^6 kilometers) from earth in 1936. { ə'dän·əs }

Adrastea [ASTRON] A small satellite of Jupiter, having an orbital radius of 80,140 miles (128,980 kilometers) and radial dimensions of 7, 6, and 5 miles (12, 10, and 8 kilometers). Also known as Jupiter XV. { ə'dras·tē·ə }

advance of the perihelion The slow rotation of the major axis of a planet's orbit in the direction of the planet's revolution, due to gravitational interactions with other planets and other effects such as those of general relativity. { əd'vans əv thə ¦per· ə¦hēl·yən }

aeon [ASTRON] A billion (10^9) years. Also spelled eon. { 'ē‚än }

A/E ratio *See* absorptivity-emissivity ratio. { ¦ā¦ē ‚rā·shō }

aerodynamic decelerator [AERO ENG] Any device made from textiles, such as a parachute, that is designed to produce drag. { ‚e·rō·dī¦nam·ik ‚de'sel·ə‚rād·ər }

aerospace ground equipment [AERO ENG] Support equipment for air and space vehicles. Abbreviated AGE. { ¦e·rō¦spās 'graund ‚kwip·mənt }

aerospace vehicle [AERO ENG] A vehicle capable of flight both within and outside the sensible atmosphere. { ¦e·rō¦spās 'vē·ə·kəl }

aerospike engine [AERO ENG] An advanced liquid-propellant rocket engine that uses an axisymmetric plug nozzle, in combination with a torus-shaped combustion chamber and a turbine exhaust system that injects the turbine drive gases into the nozzle base, to achieve a geometry that is only one-quarter the length of a conventional rocket engine, as well as automatic altitude compensation, resulting in superior low-altitude performance. { ,e·rō¦spīk 'en·jən }

aestival *See* estival. { 'es·tə·vəl }

afterbody [AERO ENG] **1.** A companion body that trails a satellite. **2.** A section or piece of a rocket or spacecraft that enters the atmosphere unprotected behind the nose cone or other body that is protected for entry. **3.** The afterpart of a vehicle. { 'af·tər,bäd·ē }

Agamemnon [ASTRON] An asteroid, one of a group of Trojan planets whose periods of revolution are approximately equal to that of Jupiter, or about 12 years. { ,ag·ə'mem·nən }

AGB *See* asymptotic giant branch.

Agena *See* β Centauri. { ə'jen·ə }

age of the moon [ASTRON] The elapsed time, usually expressed in days, since the last new moon. { 'āj əv thə 'mün }

Air Pump *See* Antlia. { 'er ,pəmp }

Aitken's formula [ASTRON] The expression used to determine the separation limit for true binary stars: log $p'' = 2.5 - 0.2m$, where p'' = limit, m = magnitude. { 'āt·kənz ¦förm·yə·lə }

Al Velorum stars *See* dwarf Cepheids. { ¦ä¦l və'lór·əm ,stärz }

Alcor [ASTRON] The star 80 Ursae Majoris. { 'al,kòr }

Aldebaran [ASTRON] A red giant star of visual magnitude 1.06, spectral classification K5-III, in the constellation Taurus; the star α Tauri. { al'deb·ə·rən }

Algenib [ASTRON] A star in the constellation Pegasus. { ,al'jen·əb }

Algol [ASTRON] An eclipsing variable star of spectral classification B8 in the constellation Perseus; the star β Persei. Also known as Demon Star. { 'al,gòl }

Algol symbiotic [ASTRON] A symbiotic star consisting of a red giant, a main-sequence star, and an accretion disk of gas from the red giant that forms around the main-sequence star and is heated by it. { 'al,gòl ,sim·bē'äd·ik }

Alioth [ASTRON] Traditional name for a second-magnitude star in the Big Dipper; the star ε Ursae Majoris. { 'al·ē,äth }

almucantar *See* parallel of altitude. { ¦al·myü¦kan·tər }

Alnilam [ASTRON] A star in the constellation Orion. { al'nil·əm }

Alpha Centauri [ASTRON] A double star, the brightest in the constellation Centaurus; apart from the sun, it is the nearest bright star to earth, about 4.3 light-years away; spectral classification G2. Also known as Rigil Kent. { ¦al·fə sen'tò·rē }

Alphonsus [ASTRON] A moon crater. { al'fän·səs }

Altair [ASTRON] A star that is 16.5 light-years from the sun; spectral type A7IV-V. Also known as α Aquilae. { al'tīr *or* 'al,ter }

Altar *See* Ara. { 'al,tär }

altitude circle *See* parallel of altitude. { 'al·tə,tüd ,sər·kəl }

Amalthea [ASTRON] The innermost known satellite of Jupiter, orbiting at a mean distance of 1.13×10^5 miles (1.82×10^5 kilometers); it has a diameter of about 150 miles (240 kilometers). Also known as Jupiter V. { ,äm·äl'thē·ə }

AM CVn star [ASTRON] A binary system consisting of two orbiting white dwarf stars which fill their respective Roche lobes, exchange mass at the point of contact, and vary in brightness due to eclipses of one white dwarf by the other. { ¦ā¦em ¦sē¦vē'en ,stär }

American Ephemeris and Nautical Almanac [ASTRON] An annual publication of the U.S. Naval Observatory containing tables of the predicted positions of various celestial bodies and other data of use to astronomers and navigators. { ə'mer·ə·kən i'fem·ə·rəs an 'nòd·i·kəl 'òl·mə,nak }

AM Herculis star *See* polar. { ¦ā¦em 'hər·kyə·ləs ¦stär }

Amor [ASTRON] An asteroid with an orbital eccentricity of 0.448 that approached to about 1×10^7 miles (1.6×10^7 kilometers) from earth. { 'ä,mòr }

Amor object [ASTRON] Any asteroid which crosses the orbit of Mars. { 'ä,mòr 'äb·jekt }

Amphitrite [ASTRON] An asteroid with a diameter of about 120 miles (200 kilometers),

3

mean distance from the sun of 2.554 astronomical units, and S-type surface composition.   { 'am·fə'trī,dē }

amplitude   [ASTRON] The range in brightness of a variable star, usually expressed in magnitudes.   { 'am·plə,tüd }

analemma   [ASTRON] A figure-eight-shaped diagram on a globe showing the declination of the sun throughout the year and also the equation of time.   { ,an·ə'lem·ə }

Ananke   [ASTRON] A small satellite of Jupiter with a diameter of about 14 miles (23 kilometers), orbiting with retrograde motion at a mean distance of 1.3 × 10⁷ miles (2.1 × 10⁷ kilometers). Also known as Jupiter XII.   { ə'naŋ·kē }

And See Andromeda.

Andr See Andromeda.

androgynous docking system   [AERO ENG] A docking system that uses the principle of reverse symmetry to allow the mechanical linkage of otherwise dissimilar spacecraft.   { an,dräj·ə·nəs 'däk·iŋ ,sis·təm }

Andromeda   [ASTRON] A constellation with a right ascension of 1 hour and a declination of 40°N. Abbreviated And; Andr.   { ,an'dräm·ə·də }

Andromeda Galaxy   [ASTRON] The spiral galaxy of type Sb nearest to the Milky Way. Also known as Andromeda Nebula.   { ,an'dräm·ə·də 'gal·ək·sē }

Andromeda Nebula See Andromeda Galaxy.   { ,an'dräm·ə·də 'ne·byə·lə }

Andromedids   [ASTRON] A meteor shower whose radiant is located near the star γ Andromedae, and which reaches its peak about November 27; associated with the Biela comet. Also known as Bielids.   { ,an'dräm·ə,didz }

anemic galaxy   [ASTRON] A spiral galaxy with unusually low surface brightness and inconspicuous spiral arms.   { ə,nēm·ik 'gal·ik·sē }

angle of commutation   [ASTRON] The difference between the celestial longitudes of the sun and a planet, as observed from the earth.   { 'aŋ·gəl əv ,käm·yə'tā·shən }

angle of vertical   [ASTRON] The angle on the celestial sphere between a given vertical circle and the prime vertical circle.   { 'aŋ·gəl əv 'vərd·ə·kəl }

angular diameter   [ASTRON] The angle subtended at the observer by a diameter of a distant spherical body which is perpendicular to the line between the observer and the center of the body.   { 'an·gyə·lər ,dī'am·əd·ər }

anisotropy   [ASTRON] The departure of the cosmic microwave radiation from equal intensity in all directions.   { ,a,nī'sä·trə·pē }

annual aberration   [ASTRON] Aberration caused by the velocity of the earth's revolution about the sun.   { 'an·yə·wəl ab·ə'rā·shən }

annual equation   [ASTRON] A variation in the moon's apparent motion caused by variations in the distance of the earth from the sun during the course of the year.   { 'an·yə·wəl i'kwā·zhən }

annual parallax   [ASTRON] The apparent displacement of a celestial body viewed from two separated observation points whose base line is the radius of the earth's orbit. Also known as heliocentric parallax.   { 'an·yə·wəl 'par·ə,laks }

annual variation   [ASTRON] The change in the right ascension and declination of a star during one year, due to the combined effect of the star's proper motion and the precession of the equinoxes.   { 'an·yə·wəl ver·ē'ā·shən }

annular eclipse   [ASTRON] An eclipse in which a thin ring of the source of light appears around the obscuring body.   { 'an·yə·lər i'klips }

anomalistic month   [ASTRON] The average period of revolution of the moon from perigee to perigee, a period of 27 days 13 hours 18 minutes 33.2 seconds.   { ə¦näm·ə¦lis·tik 'mənth }

anomalistic period   [ASTRON] The interval between two successive perigee passages of a satellite in orbit about a primary. Also known as perigee-to-perigee period.   { ə¦näm·ə¦lis·tik 'pir·ē·əd }

anomalistic year   [ASTRON] The period of one revolution of the earth about the sun from perihelion to perihelion; 365 days 6 hours 13 minutes 53.0 seconds in 1900 and increasing at the rate of 0.26 second per century.   { ə¦näm·ə¦lis·tik 'yēr }

anomaly   [ASTRON] In celestial mechanics, the angle between the radius vector to an orbiting body from its primary (the focus of the orbital ellipse) and the line of apsides of the orbit, measured in the direction of travel, from the point of closest approach to the primary (perifocus). Also known as true anomaly. { ə'näm·ə·lē }

ansae [ASTRON] **1.** The ends of the rings of Saturn, as seen from the earth. **2.** Opposing extension or knots of a celestial object, such as a planetary nebula or lenticular galaxy. { 'an·sē }

Ant *See* Antlia.

Antares [ASTRON] A red supergiant variable binary star of stellar magnitude 0.9, 520 light-years from the sun, spectral classification M1-Ib, in the constellation Scorpius; the star α Scorpii. { an'tar·ēz }

ante meridian [ASTRON] **1.** A section of the celestial meridian; it lies below the horizon, and the nadir is included. **2.** Before noon, or the period of time between midnight (0000) and noon (1200). { ¦an·tē mə¦rid·ē·ən }

anthelic arc [ASTRON] A rare type of halo phenomenon appearing in an area 180° from the sun's azimuth and at the sun's elevation. { ant'hē·lik 'ärk }

anthelion [ASTRON] A luminous white spot which occasionally appears on the parhelic circle 180° in azimuth away from the sun. Also known as counter sun; mock sun. { ant'hēl·yən }

anthropic principle [ASTRON] The assertion that the presence of intelligent life on earth places limits on the many ways the universe could have developed and could have caused the conditions of temperature that prevail today. { an'thräp·ik 'prin·sə·pəl }

anticenter [ASTRON] The direction in the sky opposite to that of the center of the Galaxy, located in the constellation Auriga. { ¦an·tē'sent·ər }

anticrepuscular rays [ASTRON] Extensions of crepuscular rays, converging toward a point 180° from the sun. { ¦an·tē·kri'pəs·kyə·lər 'rāz }

antiepicenter *See* anticenter. { ¸an·tē'ep·i¸sent·ər }

antinode [ASTRON] Either of the two points on an orbit where a line in the orbit plane, perpendicular to the line of nodes and passing through the focus, intersects the orbit. { 'an·tə¸nōd }

antisolar point [ASTRON] The point on the celestial sphere which lies directly opposite the sun from the observer, that is, on the line from the sun through the observer. { ¦an·tē¦sō·lər ¸póint }

antitail [ASTRON] A structure occasionally observed in comets that appears to extend from the coma toward the sun, and usually has the appearance of a spike. { ¦an·tē¦tāl }

Antlia [ASTRON] A constellation with a right ascension of 10 hours and declination of 35°S. Abbreviated Ant. Also known as Air Pump. { 'ant·lē·ə }

Antonadi scale [ASTRON] A scale for measuring seeing conditions, ranging from I for perfect conditions to V for very bad conditions. { ¸än·tō'näd·ē ¸skāl }

A-1 time [ASTRON] A particular atomic time scale, established by the U.S. Naval Observatory, with the origin on January 1, 1958, at zero hours Universal Time and with the unit (second) equal to 9,192,631,770 cycles of cesium at zero field. { ¦ā ¦wən ¸tīm }

apareon [ASTRON] The point on a Mars-centered orbit where a satellite is at its greatest distance from Mars. { ¸a'par·ē·ən }

apastron [ASTRON] That point of the orbit of one member of a binary star system at which the stars are farthest apart. { ¸a'pas·trən }

aphelion [ASTRON] The point on a planetary orbit farthest from the sun. { ə'fēl·yən }

aphesperian [ASTRON] The farthest point of a satellite in its orbit about Venus. { ¸a·fə'spir·ē·ən }

apoapsis [ASTRON] The point in an orbit farthest from the center of attraction. { ¦ap·ō¦ap·səs }

apocenter *See* apofocus. { 'ap·ə¸sen·tər }

apocronus [ASTRON] The farthest point of a satellite in its orbit about Saturn. Also known as aposaturnium. { ¦ap·ə¦krō·nəs }

apofocus [ASTRON] The point on an elliptic orbit at the greatest distance from the principal focus. Also known as apocenter. { ¦ap·ə¦fō·kəs }

apogalacteum [ASTRON] The point at which a celestial body is farthest from the center of the Milky Way. { ¦ap·ə·gə'lak·tē·əm }

apogee [ASTRON] That point in an orbit at which the moon or an artificial satellite is most distant from the earth; the term is sometimes loosely applied to positions of satellites of other planets. { 'ap·ə,jē }

apojove [ASTRON] The farthest point of a satellite in its orbit about Jupiter. { ¦ap·ə¦jōv }

Apollo [ASTRON] **1.** To the Greeks, the planet Mercury when it was a morning star. **2.** An asteroid with a very eccentric orbit and perihelion inside the orbit of Venus that passed about 1.8 × 10⁶ miles (3 × 10⁶ kilometers) from earth in 1932. { ə'päl·ō }

Apollo object [ASTRON] Any asteroid which crosses the earth's orbit. { ə'päl·ō 'äb·jikt }

Apollo program [AERO ENG] The scientific and technical program of the United States that involved placing men on the moon and returning them safely to earth. { ə'päl·ō ¦prō·grəm }

apolune [ASTRON] Farthest point of a satellite in an elliptic orbit about the moon. Also known as aposelene. { ¦ap·ə¦lün }

apomercurian [ASTRON] The farthest point of a satellite in its orbit about Mercury. { ¦ap·ə,mər'kyür·ē·ən }

apoplutonian [ASTRON] The farthest point of a satellite in its orbit about Pluto. { ¦ap·ə·plü'tōn·ē·ən }

apoposeidon [ASTRON] The farthest point of a satellite in its orbit about Neptune. { ¦ap·ə·pə'sīd·ən }

aposaturnium See apocronus. { ¦ap·ə·sə'tər·nē·əm }

aposelene See apolune. { ¦ap·ə·sə¦lēn }

apouranian [ASTRON] The farthest point of a satellite in its orbit about Uranus. { ¦ap·ō,yú'rän·ē·ən }

apparent [ASTRON] A term used to designate certain measured or measurable astronomic quantities to refer them to real or visible objects, such as the sun or a star. { ə'pa·rənt }

apparent horizon See horizon. { ə'pa·rənt hə'rīz·ən }

apparent libration in longitude See lunar libration. { ə'pa·rənt lī'brā·shən in 'län·jə,tüd }

apparent magnitude [ASTRON] An index of a star's brightness relative to that of the other stars; it does not take into account the difference in distance between the stars and is not an indication of the star's true luminosity. { ə'pa·rənt 'mag·nə,tüd }

apparent noon [ASTRON] Twelve o'clock apparent time, or the instant the apparent sun is over the upper branch of the meridian. { ə'pa·rənt 'nün }

apparent place See apparent position. { ə'pa·rənt 'plās }

apparent position [ASTRON] The position on the celestial sphere at which a heavenly body (or a space vehicle) would be seen from the center of the earth at a particular time. Also known as apparent place. { ə'pa·rənt pə'sish·ən }

apparent solar day [ASTRON] The duration of one rotation of the earth on its axis with respect to the apparent sun. Also known as true solar day. { ə'pa·rənt ¦sō·lər 'dā }

apparent solar time [ASTRON] Time measured by the apparent diurnal motion of the sun. Also known as apparent time; true solar time. { ə'pa·rənt ¦so·lər 'tīm }

apparent sun [ASTRON] The sun as it appears to an observer. Also known as true sun. { ə'pa·rənt 'sən }

apparent time See apparent solar time. { ə'pa·rənt 'tīm }

apparition [ASTRON] A period during which a planet, asteroid, or comet is observable, generally between two successive conjunctions of the body with the sun. { ,ap·ə'rish·ən }

applications technology satellite [AERO ENG] Any artificial satellite in the National Aeronautics and Space Administration program for the evaluation of advanced techniques and equipment for communications, meteorological, and navigation satellites. Abbreviated ATS. { ,ap·lə'kā·shənz ¦tek'näl·ə·jē ¦sad·ə,līt }

appulse [ASTRON] **1.** The near approach of one celestial body to another on the celestial sphere, as in occultation or conjunction. **2.** A penumbral eclipse of the moon. { ə'pəls }

apron [AERO ENG] A protective device specially designed to cover an area surrounding the fuel inlet on a rocket or spacecraft. { 'ā·prən }

Aps See Apus.

apse See apsis. { aps }

apsidal motion [ASTRON] The precession of the periastron of a binary system in the orbital plane of the two stars, resulting from tidal gravitational moments. { 'ap·sə· dəl 'mō·shən }

apsis [ASTRON] In celestial mechanics, either of the two orbital points nearest or farthest from the center of attraction. Also known as apse. { 'ap·səs }

Apus [ASTRON] A constellation with a right ascension of 16 hours and declination of 75°S. Abbreviated Aps. { 'ā·pəs }

Aqil See Aquila.

Aql See Aquila.

Aqr See Aquarius.

δ Aquarids [ASTRON] A meteor shower consisting of relatively long, slow-moving meteors that has its maximum around July 30 and radiant near the star δ Aquarii. { ¦del·tə 'ak·wə,ridz }

η Aquarids [ASTRON] A meteor shower associated with Halley's Comet that occurs in the first week of May with radiant near the star η Aquarii. { ¦ā·də 'ak·wə,ridz }

Aquarius [ASTRON] A constellation with a right ascension of 23 hours and declination of 15°S. Abbreviated Aqr. Also known as Water Bearer. { ə'kwer·ē·əs }

Aquila [ASTRON] A constellation with a right ascension of 20 hours and declination of 5°N. Abbreviated Aqil; Aql. { 'ak·wə·lə }

α Aquilae See Altair. { ¦al·fə 'ak·wə,lē }

Ara [ASTRON] A constellation with a right ascension of 17 hours and declination of 55°S. Also known as Altar. { 'ä·rə }

Arago distance [ASTRON] The angular distance from the antisolar point to the Arago point. { 'a·rə,gō 'dis·təns }

Arc [ASTRON] A radio source consisting of two bundles of parallel filaments adjoining the source Sagittarius A near the center of the Milky Way Galaxy. { ärk }

archeoastronomy [ASTRON] The study which attempts to reconstruct the astronomical knowledge and activity of prehistoric people and its influence on their cultures and societies. { ¦är·kē·ō·ə'strän·ə·mē }

Archer See Sagittarius. { 'är·chər }

arc jet engine [AERO ENG] An electromagnetic propulsion engine used to supply motive power for flight; hydrogen and ammonia are used as the propellant, and some plasma is formed as the result of electric-arc heating. { ¦ärk ¦jet 'en·jən }

Arcturus [ASTRON] A star that is 36 light-years from the sun; spectral classification K2IIIp. Also known as α Boötes. { ,ärk'túr·əs }

areal velocity [ASTROPHYS] In celestial mechanics, the area swept out by the radius vector per unit time. { 'er·ē·əl və'läs·əd·ē }

areocentric [ASTRON] With Mars as a center. { ,ar·ē·ō'sen·trik }

areodesy [ASTRON] Determination, by observation and measurement, of the exact positions of points on, and the figures and areas of large portions of, the surface of the planet Mars, or the shape and size of the planet Mars. { 'ar·ē·ō,des·ē }

areographic [ASTRON] Referring to positions on Mars measured in latitude from the planet's equator and in longitude from a reference meridian. { ,ar·ē·ō'graf·ik }

areography [ASTRON] The study of the surface features of Mars, or its geography. { ,ar· ē'äg·rə·fē }

areology [ASTRON] The scientific study related to the properties of Mars. { ,ar·ē'äl· ə·jē }

Ares [ASTRON] The planet Mars. { 'er,ēz }

Arethusa [ASTRON] An asteroid with a diameter of about 126 miles (210 kilometers), mean distance from the sun of 3.069 astronomical units, and C-type surface composition. { ,ar·ə'thü·zə }

Arg See Argo.

7

Argelander method   [ASTRON] A technique to estimate the brightness of variable stars; it involves estimating the difference in magnitude between the variable stars as compared to one or more stars that are invariable.   { 'är·gə,land·ər ,meth·əd }

Argo   [ASTRON] The large Ptolemy constellation; a southern constellation, now divided into four groups (Carina, Pupis, Vela, and Pyxis Nautica). Abbreviated Arg. Also known as Ship.   { 'är·gō }

argument   [ASTRON] An angle or arc, as in argument of perigee.   { 'är·gyə·mənt }

argument of latitude   [ASTRON] The angular distance measured in the orbit plane from the ascending node to the orbiting object; the sum of the argument of perigee and the true anomaly.   { 'är·gyə·mənt əv 'lad·ə,tüd }

argument of perigee   [ASTRON] The angle or arc, as seen from a focus of an elliptical orbit, from the ascending node to the closest approach of the orbiting body to the focus; the angle is measured in the orbital plane in the direction of motion of the orbiting body.   Also known as argument of perihelion.   { 'är·gyə·mənt əv 'per·ə·jē }

argument of perihelion   See argument of perigee.   { 'ar·gyə·mənt əv ¦per·ə¦hēl·yən }

ε Argus   [ASTRON] A former name of ε Carinae, a star of visual magnitude 1.74, spectral type K0.   { 'ep·sə,län 'är·gəs }

Ari   See Aries.

Ariel   [ASTRON] A satellite of the planet Uranus orbiting at a mean distance of 119,000 miles (192,000 kilometers).   { 'ar·ē·əl }

Aries   [ASTRON] A constellation with a right ascension of 3 hours and declination of 20°N. Abbreviated Ari. Also known as Ram.   { 'er,ēz }

Aristarchus   [ASTRON] A crater on the moon.   { ,ar·ə'stär·kəs }

arm population   See population I.   { 'ärm ,päp·yə,lā·shən }

Arrow   See Sagitta.   { 'ar·ō }

artificial asteroid   [AERO ENG] An object made by humans and placed in orbit about the sun.   { ¦ärd·ə¦fish·əl 'as·tə,ròid }

artificial gravity   [AERO ENG] A simulated gravity established within a space vehicle by rotation or acceleration.   { ¦ärd·ə¦fish·əl 'grav·əd·ē }

artificial satellite   [AERO ENG] Any human-made object placed in a near-periodic orbit in which it moves mainly under the gravitational influence of one celestial body, such as the earth, sun, another planet, or a planet's moon.   { ¦ärd·ə¦fish·əl 'sad·ə,līt }

ascending node   [ASTRON] Also known as northbound node. **1.** The point at which a planet, planetoid, or comet crosses to the north side of the ecliptic. **2.** The point at which a satellite crosses to the north side of the equatorial plane of its primary.   { ə'send·iŋ ¦nōd }

ascent   [AERO ENG] Motion of a craft in which the path is inclined upward with respect to the horizontal.   { ə'sent }

ashen light   [ASTRON] A faint, luminous glow sometimes observed over the right side of Venus when it is close to inferior conjunction, probably due to electrical disturbances in the ionosphere of Venus.   { 'ash·ən ¦līt }

aspect   [ASTRON] The apparent position of a celestial body relative to another; particularly, the apparent position of the moon or a planet relative to the sun.   { 'a,spekt }

association   [ASTRON] A sparsely populated grouping of very young stars that appear to have had a common origin and have not yet had time to disperse.   { ə,sō·sē'ā·shən }

A star   See A-type star.   { 'ā ,stär }

asterism   [ASTRON] A constellation or small group of stars.   { 'as·tə,riz·əm }

asteroid   [ASTRON] One of the many small celestial bodies revolving around the sun, most of the orbits being between those of Mars and Jupiter. Also known as minor planet; planetoid.   { 'as·tə,ròid }

asteroid belt   [ASTRON] The region between 2.1 and 3.5 astronomical units from the sun where most of the asteroids are found.   { 'as·tə,ròid ,belt }

astral   [ASTRON] Characteristic of a specific star or stars; stellar is the accepted term.   { 'as·trəl }

astre fictif   [ASTRON] Any of several fictitious stars assumed to move along the celestial equator at uniform rates corresponding to the speeds of the several harmonic constituents of the tide-producing force. { 'as·tər 'fik,tif }

astro- [ASTRON] A prefix meaning star or stars and, by extension, sometimes used as the equivalent of celestial, as in astronautics. { 'as·trō }

astrochemistry [ASTRON] The science that applies the principles of chemistry to matter in space. { ¦as·trō'kem·ə·strē }

astrochronology [ASTRON] The use of stellar phenomena in chronology. { ¦as·trō· krə'näl·ə·jē }

astrodome [AERO ENG] A transparent dome in the fuselage or body of an aircraft or spacecraft intended primarily to permit taking celestial observations in navigating. Also known as astral dome; navigation dome. { 'as·trō,dōm }

astrodynamics [AERO ENG] The practical application of celestial mechanics, astroballistics, propulsion theory, and allied fields to the problem of planning and directing the trajectories of space vehicles. [ASTROPHYS] The dynamics of celestial objects. { ,as·trō·dī'nam·iks }

astrogeology [ASTRON] The science that applies the principles of geology, geochemistry, and geophysics to the moon and planets other than the earth. { ,as·trō,jē'äl· ə·jē }

astrograph [ASTRON] A telescope designed to be used exclusively for astronomical photography. { 'as·trō,graf }

astrographic position See astrometric position. { ¦as·trō¦graf·ik pə'zish·ən }

astrograph mean time [ASTRON] A form of mean time, used in setting an astrograph; mean-time setting of 1200 occurs when the local hour angle of Aries is 0°. { 'as· trō,graf ¦mēn 'tīm }

astrometric binary star [ASTRON] A binary star that may be distinguished from a single star only from the variable proper motion of one of its components. { ¦as·trə¦me· trik 'bī,ner·ē 'stär }

astrometric position [ASTRON] The position of a heavenly body or space vehicle on the celestial sphere corrected for aberration but not for planetary aberration. Also known as astrographic position. { ¦as·trə¦me·trik pə'zish·ən }

astrometry [ASTRON] The branch of astronomy dealing with the geometrical relations of the celestial bodies and their real and apparent motions. { ə'sträm·ə·trē }

astronaut [AERO ENG] In United States terminology, a person who rides in a space vehicle. { 'as·trə,nȯt }

astronautical engineering [AERO ENG] The engineering aspects of flight in space. { ¦as·trə¦nȯd·ə·kəl ,en·jə'nir·iŋ }

astronautics [AERO ENG] **1.** The art, skill, or activity of operating spacecraft. **2.** The science of space flight. { ,as·trə'nȯd·iks }

astronomic See astronomical. { ,as·trə'näm·ik }

astronomical [ASTRON] Of or pertaining to astronomy or to observations of the celestial bodies. Also known as astronomic. { ,as·trə'näm·ə·kəl }

astronomical almanac [ASTRON] A publication giving the tables of coordinates of a number of celestial bodies at a number of specific times during a given period. { ,as·trə'näm·ə·kəl 'ȯl·mə,nak }

astronomical atlas [ASTRON] A set of maps of celestial phenomena, often developed in conjunction with an astronomical catalog, and providing a clear picture of the spatial relations between the phenomena. { ,as·trə'näm·ə·kəl 'at·ləs }

astronomical catalogue [ASTRON] A list or enumeration of astronomical data, generally ordered by increasing right ascension of the objects listed. { 'as·trə'näm·ə·kəl 'kad·əl,äg }

astronomical constants [ASTROPHYS] The elements of the orbits of the bodies of the solar system, their masses relative to the sun, their size, shape, orientation, rotation, and inner constitution, and the velocity of light. { ,as·trə'näm·ə·kəl 'kän·stəns }

astronomical coordinate system [ASTRON] Any system of spherical coordinates serving to locate astronomical objects on the celestial sphere. { ,as·trə'näm·ə·kəl ,kō'ȯrd· ə·nət ,sis·təm }

astronomical date [ASTRON] Designation of epoch by year, month, day, and decimal fraction. { ˌas·trə'näm·ə·kəl 'dāt }

astronomical day [ASTRON] A mean solar day beginning at mean noon, 12 hours later than the beginning of the civil day of the same date; astronomers now generally use the civil day. { ˌas·trə'näm·ə·kəl 'dā }

astronomical distance [ASTRON] The distance of a celestial body expressed in units such as the light-year, astronomical unit, and parsec. { ˌas·trə'näm·ə·kəl 'dis·təns }

astronomical eclipse See eclipse. { ˌas·trə'näm·ə·kəl i'klips }

astronomical ephemeris See ephemeris. { ˌas·trə'näm·ə·kəl i'fem·ə·rəs }

astronomical nutation [ASTRON] A small periodic motion of the celestial pole of celestial bodies, including the earth, with respect to the pole of the ecliptic. { ˌas·trə'näm·ə·kəl nü'tā·shən }

astronomical observatory [ASTRON] A building designed and equipped for making observations of astronomical phenomena. { ˌas·trə'näm·ə·kəl əb'zər·və,tör·ē }

astronomical scintillation [ASTROPHYS] Any scintillation phenomena, such as irregular oscillatory motion, variation of intensity, and color fluctuation, observed in the light emanating from an extraterrestrial source. Also known as stellar scintillation. { ˌas·trə'näm·ə·kəl sint·əl'ā·shən }

astronomical seeing See seeing. { ˌas·trə'näm·ə·kəl 'sē·iŋ }

astronomical time [ASTRON] Solar time in an astronomical day that begins at noon. { ˌas·trə'näm·ə·kəl 'tīm }

astronomical triangle [ASTRON] A spherical triangle on the celestial sphere. { ˌas·trə'näm·ə·kəl 'trī,aŋ·gəl }

astronomical twilight [ASTRON] The period of incomplete darkness when the center of the sun is more than 6° but not more than 18° below the celestial horizon. { ˌas·trə'näm·ə·kəl 'twī,līt }

astronomical unit [ASTRON] Abbreviated AU. **1.** A measure for distance within the solar system equal to the mean distance between earth and sun, that is, about 92,956,000 kilometers (149,598,000 kilometers). **2.** The semimajor axis of the elliptical orbit of earth. { ˌas·trə'näm·ə·kəl 'yü·nət }

astronomical year See tropical year. { ˌas·trə'näm·ə·kəl 'yir }

astrophysics [ASTRON] A branch of astronomy that treats of the physical properties of celestial bodies, such as luminosity, size, mass, density, temperature, and chemical composition, and with their origin and evaluation. { ˌas·trō'fiz·iks }

asymptotic giant branch [ASTRON] A grouping of stars on the Hertzsprung-Russell diagram that is roughly asymptotic to the giant branch; it represents a later stage in giant-star evolution in which hydrogen-fusing and helium-fusing shells surround a core in which both hydrogen fusion and helium fusion are exhausted. Abbreviated AGB. { ˌa,sim'täd·ik ,jī·ənt 'branch }

Aten [ASTRON] The first asteroid found to have a period less than that of the earth, 346.93 days, with an orbital eccentricity of 0.19. { 'ä,ten }

Aten asteroid [ASTRON] An asteroid whose period is less than that of the earth. { 'ä,ten 'as·tə,rȯid }

Atlantic standard time See Atlantic time. { ət'lan·tik 'stan·dərd ,tīm }

Atlantic time [ASTRON] A time zone; the fourth zone west of Greenwich. Also known as Atlantic standard time. { ət'lan·tik 'tīm }

Atlas [ASTRON] The innermost known satellite of Saturn, which orbits at a distance of 85 × 10³ miles (137 × 10³ kilometers), just outside the A ring, and has an irregular shape with an average diameter of 20 miles (30 kilometers). { 'at·ləs }

atmospheric braking [AERO ENG] **1.** Slowing down an object entering the atmosphere of the earth or other planet from space by using the drag exerted by air or other gas particles in the atmosphere. **2.** The action of the drag so exerted. { 'at·mə'sfir·ik 'brāk·iŋ }

atmospheric entry [AERO ENG] The penetration of any planetary atmosphere by any object from outer space; specifically, the penetration of the earth's atmosphere by a crewed or uncrewed capsule or spacecraft. { 'at·mə'sfir·ik 'en·trē }

atomic nucleus See nucleus. { ə'täm·ik 'nü·klē·əs }

atomic rocket   [AERO ENG] A rocket propelled by an engine in which the energy for the jetstream is to be generated by nuclear fission or fusion.   Also known as nuclear rocket.   { ə'täm·ik 'räk·ət }

attitude   [AERO ENG] The position or orientation of an aircraft, spacecraft, and so on, either in motion or at rest, as determined by the relationship between its axes and some reference line or plane or some fixed system of reference axes.   { 'ad·ə,tüd }

attitude control   [AERO ENG] **1.** The regulation of the attitude of an aircraft, spacecraft, and so on.   **2.** A device or system that automatically regulates and corrects attitude, especially of a pilotless vehicle.   { 'ad·ə,tüd kən,trōl }

attitude gyro   [AERO ENG] Also known as attitude indicator.   **1.** A gyro-operated flight instrument that indicates the attitude of an aircraft or spacecraft with respect to a reference coordinate system.   **2.** Any gyro-operated instrument that indicates attitude.   { 'ad·ə,tüd ,jī·rō }

attitude jet   [AERO ENG] **1.** A stream of gas from a jet used to correct or alter the attitude of a flying body either in the atmosphere or in space.   **2.** The nozzle that directs this jetstream.   { 'ad·ə,tüd ,jet }

A-type star   [ASTRON] In star classification based on spectral characteristics, the type of star in whose spectrum the hydrogen absorption lines are at a maximum.   Also known as A star.   { 'ā,tīp ,stär }

AU *See* astronomical unit.

Auger shower   [ASTRON] A very large cosmic-ray shower.   Also known as extensive air shower.   { ō'zhā ¦shaú·ər }

augmentation   [ASTRON] The apparent increase in the semidiameter of a celestial body, as observed from the earth, as the body's altitude (angular distance above the horizon) increases, due to the reduced distance from the observer; used principally in reference to the moon.   { ,óg·mən'tā·shən }

Aur *See* Auriga.

Auri *See* Auriga.

Auriga   [ASTRON] A constellation with a right ascension of 6 hours and declination of 40°N.   Abbreviated Aur; Auri.   { ò'rī·gə }

Aurora   [ASTRON] An asteroid with a diameter of about 132 miles (220 kilometers), mean distance from the sun of 3.153 astronomical units, and C-type surface composition.   { ə'ròr·ə }

auroral   [ASTRON] The period of dusk before sunrise.   { ə'ròr·əl }

autumn   [ASTRON] The season of the year which is the transition period from summer to winter, occurring as the sun approaches the winter solstice; beginning is marked by the autumnal equinox.   Also known as fall.   { 'òd·əm }

autumnal   [ASTRON] Pertaining to the season autumn.   { ò'təm·nəl }

autumnal equinox   [ASTRON] The point on the celestial sphere at which the sun's rays at noon are 90° above the horizon at the Equator, or at an angle of 90° with the earth's axis, and neither North nor South Pole is inclined to the sun; occurs in the Northern Hemisphere on approximately September 23 and marks the beginning of autumn.   Also known as first point of Libra.   { ò'təm·nəl 'ē·kwə,näks }

auxiliary circle   [ASTRON] In celestial mechanics, a circumscribing circle to an orbital ellipse with radius *a*, the semimajor axis.   { òg'zil·yə·rē 'sər·kəl }

auxiliary fluid ignition   [AERO ENG] A method of ignition of a liquid-propellant rocket engine in which a liquid that is hypergolic with either the fuel or the oxidizer is injected into the combustion chamber to initiate combustion.   { òg'zil·yə·rē 'flü·əd ig'nish·ən }

auxiliary power unit   [AERO ENG] A power unit carried on an aircraft or spacecraft which can be used in addition to the main sources of power.   Abbreviated APU.   { òg'zil·yə·rē 'paú·ər ,yü·nət }

azimuth   [ASTRON] Horizontal direction of a celestial point from a terrestrial point, expressed as the angular distance from a reference direction, usually measured from 0° at the reference direction clockwise through 360°.   { 'az·ə·məth }

azimuth error   [ASTRON] The angle by which the east-west axis of a transit telescope deviates from being perpendicular to the plane of the meridian.   { 'az·ə·məth ,er·ər }

azimuth tables [ASTRON] Publications providing tabulated azimuths or azimuth angles of celestial bodies for various combinations of declination, altitude, and hour angle; great-circle course angles can also be obtained by substitution of values. { 'az·ə·məth ‚tā·bəlz }

B

Baade's window [ASTRON] An unusually transparent region about 4° from the galactic center. { 'bä·dəz ,win·dō }

Babcock magnetograph [ASTRON] An instrument used to measure weak magnetic fields on the sun. { 'bab,käk mag'ned·ə,graf }

backout [AERO ENG] An undoing of previous steps during a countdown, usually in reverse order. { 'bak,aút }

Baily's beads [ASTRON] Bright points of sunlight appearing around the edge of the moon just before and after the central phase of a total solar eclipse. { 'bāl,ēz ¦bēdz }

balance [AERO ENG] **1.** The equilibrium attained by an aircraft, rocket, or the like when forces and moments are acting upon it so as to produce steady flight, especially without rotation about its axes. **2.** The equilibrium about any specified axis that counterbalances something, especially on an aircraft control surface, such as a weight installed forward of the hinge axis to counterbalance the surface aft of the hinge axis. { 'bal·əns }

Balance See Libra. { 'bal·əns }

balloon [AERO ENG] A nonporous, flexible spherical bag, inflated with a gas such as helium that is lighter than air, so that it will rise and float in the atmosphere; a large-capacity balloon can be used to lift a payload suspended from it. { bə'lün }

balloon astronomy [ASTRON] The observation of celestial objects from instruments mounted on balloons and carried to altitudes up to 18 miles (30 kilometers), to detect electromagnetic radiation at wavelengths which do not penetrate to the earth's surface. { bə'lün ə'strän·ə·mē }

balloon-type rocket [AERO ENG] A liquid-fuel rocket, such as the Atlas, that requires the pressure of its propellants (or other gases) within it to give it structural integrity. { bə'lün ¦tīp 'räk·ət }

Bamberga [ASTRON] An asteroid with a diameter of about 129 miles (215 kilometers), mean distance from the sun of 2.686 astronomical units, and C-type surface composition. { 'bäm,bər·gə }

barium star [ASTRON] A peculiar, low-velocity, strong-lined red giant or subgiant star of spectral type G, K, or M, whose spectrum has anomalously strong lines of barium, sometimes with strong bands of methyldadyne (CH), molecular carbon (C_2), and cyanogen radical (CN). { 'bar·ē·əm ,stär }

Barnard's loop [ASTRON] A large emission nebula, about 10° by 140° in size, around the central portion of Orion, that consists of an expanding shell of gas that probably originated in a supernova. { 'bär·nərdz ,lüp }

Barnard's star [ASTRON] A star 6.1 light-years away from earth, of visual magnitude 9.5 and proper motion of 10.31 seconds of arc annually. { 'bär·nərdz ¦stär }

barred spiral galaxy [ASTRON] A spiral galaxy whose spiral arms originate at the ends of a bar-shaped structure centered at the nucleus of the galaxy. { ¦bärd ¦spī·rəl 'gal·ik·sē }

barycenter [ASTRON] The center of gravity of the earth-moon system. { 'bar·ə,sen·tər }

barycentric element [ASTROPHYS] An orbital element referred to the center of mass of the solar system. { ,bar·ə'sen·trik 'el·ə·mənt }

baryon-to-photon ratio [ASTRON] The estimated ratio of the number of baryons (mostly protons and neutrons) to photons (mostly in the cosmic microwave radiation) in the universe. { 'bär·ē,än tə 'fō,tän 'rā·shō }

Bautz-Morgan classification [ASTRON] A classification of clusters of galaxies into three categories, ranging from type I in which the cluster contains a supergiant elliptical galaxy, to type II in which the cluster contains no member that is significantly brighter than the general bright population. { 'baüts 'mȯr·gən ,klas·ə·fə,kā·shən }

Bayer letter [ASTRON] The Greek (or Roman) letter used in a Bayer name. { 'bī·ər ,led·ər }

Bayer name [ASTRON] The Greek (or Roman) letter and the possessive form of the Latin name of a constellation, used as a star name; examples are α Cygni (Deneb), β Orionis (Rigel), and η Ursae Majoris (Alkaid). { 'bī·ər ,nām }

Bayer's constellations [ASTRON] Thirteen constellations in the southern hemisphere named by J. Bayer. { 'bī·ərz ,kan·stə'lā·shənz }

Bear Driver See Boötes. { 'ber ,drīv·ər }

beat Cepheid [ASTRON] A dwarf Cepheid that displays two or more nearly identical pulsation periods, resulting in periodic amplitude fluctuations in its light curve. { 'bēt 'sef·ē·əd }

Becklin-Neugebauer object [ASTRON] A compact source of infrared radiation in the Orion Nebula, probably a collapsing protostar of large mass. Abbreviated BN object. { 'bek·lin 'nȯi·gə,baü·ər ,äb,jekt }

Beehive See Praesepe. { 'bē,hīv }

Belinda [ASTRON] A satellite of Uranus orbiting at a mean distance of 46,760 miles (75,260 kilometers) with a period of 15 hours, and with a diameter of about 42 miles (68 kilometers). { bə'lin·də }

Bellatrix [ASTRON] A bluish-white star of stellar magnitude 1.7, spectral classification B2-III, in the constellation Orion; the star γ Orionis. { bə'lā·triks }

Berenice's Hair See Coma Berenices. { ,ber·ə'nē·səz 'her }

Besselian elements [ASTRON] Data on a solar eclipse, giving, for selected times, the coordinates of the axis of the moon's shadow with respect to the fundamental plane, and the radii of umbra and penumbra in that plane; the data allow one to derive local circumstances of the eclipse at any point on the earth's surface. { bə'sel·yən ¦el·ə·mənts }

Besselian star numbers [ASTRON] Constants used in the reduction of a mean position of a star to an apparent position; used to account for short-term variations in precession, nutation, aberration, and parallax. { bə'sel·yən 'stär ,nəm·bərz }

Besselian year See fictitious year. { bə'sel·yən ,yir }

Be star [ASTRON] A star of spectral type B in the Draper catalog that has emission lines indicating mass loss and a surrounding gaseous shell. { ¦bē¦ē ,stär }

beta [ASTRON] For dust grains ejected from the nucleus of a comet, the ratio of the radiation pressure force to the solar gravitational force. { 'bād·ə }

beta Canis Majoris stars See beta Cephei stars. { 'bād·ə 'kan·əs mə'jȯr·əs ,stärz }

beta Centauri [ASTRON] A first-magnitude navigational star in the constellation Centaurus; 200 light-years from the sun; spectral classification B0. Also known as Agena; Hadar. { ¦bā·də sen'tȯ·rē }

beta Cephei stars [ASTRON] A class of pulsating variables lying above the upper main sequence with short periods of $3^1/_2$–6 hours, spectral classes B0 to B3, and doubly periodic light curves. Also known as beta Canis Majoris stars. { 'bād·ə 'sef·ē,ī ,stärz }

Betelgeuse [ASTRON] An orange-red giant star of stellar magnitude 0.1–1.2, 650 light-years from the sun, spectral classification M2-Iab, in the constellation Orion; the star α Orionis. { 'bed·əl,jüs }

Bianca [ASTRON] A satellite of Uranus orbiting at a mean distance of 36,760 miles (59,160 kilometers) with a period of 10 hours 27 minutes, and with a diameter of about 27 miles (44 kilometers). { bē'äŋk·ə }

Bianchi cosmology [ASTRON] A model of the universe which is homogeneous but not necessarily isotropic. { bē'aŋ·kē käz'mäl·ə·jē }

Biela Comet [ASTRON] A comet seen in 1852 at one perihelion passage; presumed to have separated into two bodies. Also known as Comet Biela. { 'bē·lä ˌkäm·ət }

Bielids See Andromedids. { 'bēˌlidz }

big bang theory [ASTRON] A theory of the origin and evolution of the universe which holds that approximately 1.4×10^{10} years ago all the matter in the universe was packed into a small agglomeration of extremely high density and temperature which exploded, sending matter in all directions and giving rise to the expanding universe. Also known as superdense theory. { ¦big 'baŋ ˌthē·ə·rē }

big crunch [ASTRON] A singularity at the origin of a black hole into which all the matter and radiation in a closed universe would eventually collapse. Also known as gnab gib. { ¦big ¦krənch }

Big Dipper [ASTRON] A group of stars that is part of the constellation Ursa Major. Also known as Charles' wain. { ¦big 'dip·ər }

Big Four [ASTRON] A group of large asteroids including Ceres, Pallas, Vesta, and Juno, the first four that were discovered. { ¦big ¦fór }

binary pulsar [ASTRON] A pulsar which forms one component of a binary star. { 'bīn·ə·rē 'pəlˌsär }

binary star [ASTRON] A pair of stars located sufficiently near each other in space to be connected by the bond of mutual gravitational attraction, compelling them to describe an orbit around their common center of gravity. Also known as binary system. { 'bīn·ə·rē 'stär }

binary system See binary star. { 'bīn·ə·rē 'sis·təm }

biosatellite [AERO ENG] An artificial satellite designed to contain and support humans, animals, or other living material in a reasonably normal manner for a period of time and to return safely to earth. { ¦bī·o'sadˌəlˌīt }

bipolar nebula [ASTRON] A nebula consisting of two relatively symmetrical bright lobes with a star between them. { bī͟ˌpōl·ər 'neb·yə·lə }

Blaauw mechanism [ASTRON] An explanation for the disruption of a binary system as being due to the decrease in the gravitational binding force when a shell of gas ejected by the primary component overtakes the secondary. { 'blō ˌmek·əˌniz·əm }

black drop [ASTRON] As seen through a telescope, an apparent dark elongation of the image of Venus or Mercury when the planets' images are at the sun's limb. { ¦blak 'dräp }

black dwarf See brown dwarf. { ¦blak 'dwórf }

blast chamber [AERO ENG] A combustion chamber, especially in a gas-turbine, jet, or rocket engine. { 'blast ˌchām·bər }

blast deflector [AERO ENG] A device used to divert the exhaust of a rocket fired from a vertical position. { 'blast di'flek·tər }

blast-off [AERO ENG] The takeoff of a rocket or missile. { 'blastˌóf }

blazar [ASTRON] A type of quasar whose light exhibits strong optical polarization and large variability. { 'bläˌzär }

BL Herculis stars [ASTRON] W Virginis stars of relatively low luminosity and mass. { ¦beˌel 'hər·kyə·ləs ˌstärz }

BL Lacertae objects [ASTRON] A class of extragalactic sources of extremely intense, highly variable electromagnetic radiation which are related to quasars but have a featureless optical spectrum, and display strong optical polarization and a radio spectrum that increases in intensity at shorter wavelengths. { ¦bēˌel lə'ser·tē ˌäb·jiks }

blowoff [AERO ENG] The action of applying an explosive force and separating a package section away from the remaining part of a rocket vehicle or reentry body, usually to retrieve an instrument or to obtain a record made during early flight. { 'blōˌóf }

blue band [ASTRON] A dark band which appears around the polar caps of Mars as they shrink during the spring and early summer. { ¦blü 'band }

blue edge [ASTRON] The curve on the Hertzsprung-Russell diagram given by the maximum temperature, as a function of luminosity, at which a star of specified composition is unstable against small-amplitude pulsations. { 'blü 'ej }

blue flash See green flash. { 'blü ˌflash }

blue-green flame *See* green flash. { ¦blü¦grēn 'flām }

blue haze |ASTRON| A condition of the Martian atmosphere that sometimes causes it to be opaque to radiation near the blue end of the visible spectrum. { 'blü 'hāz }

blue shift |ASTRON| A displacement of lines in the spectrum of a celestial object toward shorter wavelengths, indicating motion of the object toward the observer. { 'blü ‚shift }

blue star |ASTRON| A star of spectral type O, B, A, or F according to the Draper catalog. { 'blü ‚stär }

blue straggler star |ASTRON| A member of a star cluster that lies above the turnoff point of the cluster's Hertzsprung-Russell diagram, and lies near the main sequence. { ¦blü 'strag·lər ‚stär }

bluff body |AERO ENG| A body having a broad, flattened front, as in some reentry vehicles. { ¦bləf ¦bäd·ē }

BN object *See* Becklin-Neugebauer object. { ¦bē'en ‚äb·jəkt }

boattail |AERO ENG| Of an elongated body such as a rocket, the rear portion having decreasing cross-sectional area. { 'bōt‚tāl }

Bode's law |ASTRON| An empirical law giving mean distances of planets to the sun by the formula $a = 0.4 + 0.3 \times 2^n$, where a is in astronomical units and n equals $-\infty$ for Mercury, 0 for Venus, 1 for Earth, and so on; the asteroids are included as planets. Also known as Titius-Bode law. { 'bōdz ‚lȯ }

body |AERO ENG| **1.** The main part or main central portion of an airplane, airship, rocket, or the like; a fuselage or hull. **2.** Any fabrication, structure, or other material form, especially one aerodynamically or ballistically designed; for example, an airfoil is a body designed to produce an aerodynamic reaction. { 'bäd·ē }

boiling |ASTRON| The telescopic appearance of the limbs of the sun and planets when the earth's atmosphere is turbulent, characterized by a constant rippling motion and lack of a clearly defined edge. { 'bȯil·iŋ }

bolide |ASTRON| A brilliant meteor, especially one which explodes; a detonating fireball meteor. { 'bō‚līd }

bolometric correction |ASTRON| The difference between the bolometric and visual magnitude. { ¦bō·lə¦me·trik kə'rek·shən }

bolometric magnitude |ASTRON| The magnitude of a celestial object, as calculated from the total amount of radiation received from the object at all wavelengths. { ¦bō·lə¦me·trik 'mag·nə‚tüd }

Boo *See* Boötes.

boost |AERO ENG| **1.** An auxiliary means of propulsion such as by a booster. **2.** To supercharge. **3.** To launch or push along during a portion of a flight. **4.** *See* boost pressure. { büst }

booster *See* booster engine; booster rocket; launch vehicle. { 'büs·tər }

booster engine |AERO ENG| An engine, especially a booster rocket, that adds its thrust to the thrust of the sustainer engine. Also known as booster. { 'büs·tər ‚en·jən }

booster rocket |AERO ENG| Also known as booster. **1.** A rocket motor, either solid- or liquid-fueled, that assists the normal propulsive system or sustainer engine of a rocket or aeronautical vehicle in some phase of its flight. **2.** A rocket used to set a vehicle in motion before another engine takes over. { 'büs·tər ‚räk·ət }

boost-glide vehicle |AERO ENG| An air vehicle capable of aerodynamic lift which is projected to an extreme altitude by reaction propulsion and then coasts down with little or no propulsion, gliding to increase its range when it reenters the sensible atmosphere. { ¦büst ¦glīd 'vē·ə·kəl }

Boot *See* Boötes. { büt }

Boötes |ASTRON| A constellation which lies south and east of Ursa Major; the star Arcturus is a member of the group. Abbreviated Boo; Boot. Also known as Bear Driver. { bō'ō‚tēz }

α Boötis *See* Arcturus. { ¦al·fə bō'ō·təs }

bootstrap process |AERO ENG| A self-generating or self-sustaining process; specifically, the operation of liquid-propellant rocket engines in which, during main-stage operation, the gas generator is fed by the main propellants pumped by the turbopump, and the turbopump in turn is driven by hot gases from the gas generator system. { 'büt,strap ,präs·əs }

bowl crater [ASTRON] A type of lunar crater whose interior cross section is a smooth curve, with no flat floor. { 'bōl ,krād·ər }

bowshock [ASTROPHYS] The shock wave set up by the interaction of the supersonic solar wind with a planet's magnetic field. { 'baŭ,shäk }

Bradley aberration [ASTRON] Stellar aberration with a maximum of 20.5 seconds of arc; can be used to compute an approximate velocity for light. { 'brad·lē ab·ə'rā·shən }

braking ellipses [AERO ENG] A series of ellipses, decreasing in size due to aerodynamic drag, followed by a spacecraft in entering a planetary atmosphere. { 'bra·kiŋ i'lip,sēz }

brennschluss [AERO ENG] **1.** The cessation of burning in a rocket, resulting from consumption of the propellants, from deliberate shutoff, or from other cause. **2.** The time at which this cessation occurs. { 'bren,shlùs }

bright diffuse nebula [ASTRON] A nebula which is illuminated by the action of embedded or nearby stars. { ¦brīt dəǀfyüs 'neb·yə·lə }

bright points [ASTRON] Relatively small regions on the sun, distributed uniformly over the solar disk, from which there is increased x-ray and ultraviolet emission, having lifetimes on the order of 8 hours. { 'brīt ,póins }

bright rim structures [ASTRON] Bright edges exhibited by many diffuse-emission nebulae, usually on the side facing the exciting star. { 'brīt ,rim ,strək·chərz }

bright stars catalog [ASTRON] A catalog of stars brighter than 6.5 magnitude, giving positions, motions, parallaxes, and spectral classes. { ¦brīt ¦stärz 'kad·ə,läg }

brown dwarf [ASTRON] A starlike body whose mass is too small (less than about 8% that of the sun) to sustain nuclear reactions in its core. Also known as black dwarf; failed star; infrared dwarf; lilliputian star; substellar object; super-Jupiter. { ¦braün ¦dwörf }

B star *See* B-type star. { 'bē ,stär }

B Tauri [ASTRON] A daytime meteor shower that occurs at the end of June and has its radiant near the star. { ¦bē 'tór·ē }

B-type star [ASTRON] A type in a classification based on stellar spectral characteristics; has strong HeI absorption. Also known as B star. { 'bē ,tīp ,stär }

bulkhead [AERO ENG] A wall, partition, or such in a rocket, spacecraft, airplane fuselage, or similar structure, at right angles to the longitudinal axis of the structure and serving to strengthen, divide, or help give shape to the structure. { 'bəlk,hed }

Bull *See* Taurus. { bəl }

bump Cepheid [ASTRON] A Cepheid variable star with a period of 5−15 days that displays a prominent secondary maximum (bump) in its light and velocity curves. { 'bəmp 'sef·ē·əd }

burning-rate constant [AERO ENG] A constant, related to initial grain temperature, used in calculating the burning rate of a rocket propellant grain. { 'bər·niŋ ,rāt ,kän·stənt }

burnout [AERO ENG] **1.** An act or instance of fuel or oxidant depletion or of depletion of both at once. **2.** The time at which this depletion occurs. **3.** The point on a rocket trajectory at which this depletion occurs. { 'bərn,aùt }

burnout velocity [AERO ENG] The velocity of a rocket at the time when depletion of the fuel or oxidant occurs. Also known as all-burnt velocity; burnt velocity. { 'bərn,aùt və'läs·əd·ē }

burst disk [AERO ENG] A diaphragm designed to burst at a predetermined pressure differential; sometimes used as a valve, for example, in a liquid-propellant line in a rocket. Also known as rupture disk. { 'bərst ,disk }

burster [ASTRON] A celestial source of radiation, such as x-rays or gamma rays, that is very intense for brief periods of time and whose nature has not yet been established. { 'bər·stər }

bus [AERO ENG] A spacecraft or missile that is designed to carry one or more separable devices, such as probes or warheads. { bəs }

C

Cae *See* Caelum.

Caelum [ASTRON] A southern constellation, right ascension 5 hours, declination 40°S. Abbreviated Cae. Also known as Chisel. { 'sē·ləm }

calcium star [ASTRON] A term sometimes used to denote a star of spectral class F, · which has prominent absorption bands of calcium. { 'kal·se·əm ˌstär }

Caldwell catalog [ASTRON] A catalog of star clusters, nebulae, and galaxies for the use of amateur observers, whose objects are easy to locate with a small telescope but are not included in Messier's Catalog. { 'kòl,dwel ¦kad·ə,läg }

Caldwell number [ASTRON] A number by which star clusters, nebulae, and galaxies are listed in the Caldwell Catalog, arranged in order of declination; for example, the Hyades are C41. { 'kòl,dwel ¦nəm·bər }

calendar [ASTRON] A system for everyday use in which time is divided into days and longer periods, such as weeks, months, and years, and a definite order for these periods and a correspondence between them are established. { 'kal·ən·dər }

calendar day [ASTRON] The period from midnight to midnight; it is 24 hours of mean solar time in length and coincides with the civil day. { 'kal·ən·dər ˌdā }

calendar month [ASTRON] The month of the calendar, varying from 28 to 31 days in length. { 'kal·ən·dər ˌmənth }

calendar year [ASTRON] The year in the Gregorian calendar, common years having 365 days and leap years 366. Also known as civil year. { 'kal·ən·dər ˌyir }

Caliban [ASTRON] A small satellite of Uranus in a retrograde orbit with a mean distance of 4,475,000 miles (7,200,000 kilometers), eccentricity of 0.081, and sidereal period of 1.59 years. { 'kal·ə,ban }

Callipic cycle [ASTRON] Four Metonic cycles, or 76 years. { kə'lip·ik ˌsī·kəl }

Callisto [ASTRON] A satellite of Jupiter orbiting at a mean distance of 1,884,000 kilometers. Also known as Jupiter IV. { kə'lis,tō }

Caloris Basin [ASTRON] A large depression on Mercury, about 1300 kilometers in diameter. { kə'lòr·əs 'bā·sən }

Calypso [ASTRON] A small, irregularly shaped satellite of Saturn that librates about the leading Lagrangian point of Tethys's orbit. { kə'lip·sō }

Cam *See* Camelopardalis.

Camelopardalis [ASTRON] Latin name for the Giraffe constellation of the northern hemisphere. Abbreviated Cam; Caml. Also known as Camelopardus; Giraffe. { ka,mel·ə'pärd·əl·əs }

Camelopardus *See* Camelopardalis. { ka,mel·ə'pär·dəs }

Camilla [ASTRON] An asteroid with a diameter of about 220 kilometers, mean distance from the sun of 3.49 astronomical units, and C-type surface composition. { kə'mil·ə }

Caml *See* Camelopardalis. { 'kam·əl }

Canc *See* Cancer.

Cancer [ASTRON] A constellation with right ascension 9 hours, declination 20°N. Abbreviated Canc. Also known as Crab. { 'kan·sər }

Canes Venatici [ASTRON] A northern constellation with right ascension 13 hours, declination 40°N, between Ursa Major and Boötes. Abbreviated CVn. Also known as Hunting Dogs. { 'kä,nēz ·və'nad·ə,sē }

Canes Venatici I cloud   [ASTRON]   A relatively nearby, loosely clustered group of galaxies consisting chiefly of late-type spirals and irregular galaxies, with recession velocities near 220 miles (350 kilometers) per second.   { 'kā¦nēz və'nad·ə¸sē 'wən ¸klaüd }

Canis Major   [ASTRON]   A constellation with right ascension 7 hours, declination 20°S. Abbreviated CMa.   Also known as Greater Dog.   { ¸kā·nəs 'mā·jər }

α **Canis Majoris**   See Sirius.   { ¦al·fə ¸kā·nəs 'mā·jər·is }

ε **Canis Majoris**   See Adhara.   { ¦ed·ə ¸kā·nəs 'mā·jər·is }

Canis Minor   [ASTRON]   A constellation with right ascension 8 hours, declination 5°N. Abbreviated CMi.   Also known as Lesser Dog.   { ¸kā·nəs 'mī·nər }

α **Canis Minoris**   See Procyon.   { ¦al·fə ¸kā·nəs 'mī·nər·is }

canonical change   [ASTRON]   A periodic change in one of the components of the orbit of a celestial object.   { kə'nän·ə·kəl 'chänj }

canonical time unit   [ASTRON]   For geocentric orbits, the time required by a hypothetical satellite to move one radian in a circular orbit of the earth's equatorial radius, that is, 13.447052 minutes.   { kə'nän·ə·kəl 'tīm ¸yü·nət }

Canopus   [ASTRON]   A star that is 180 light-years from the sun; spectral classification F0Ia.   Also known as α Carinae.   { kə'nō·pəs }

Cap   See Capricornus.

Capella   [ASTRON]   A star that is 45 light-years from the sun; spectral classification G0IIIp.   Also known as α Aurigae.   { kə'pel·ə }

Capricornus   [ASTRON]   A constellation with right ascension 21 hours, declination 20°S. Abbreviated Cap.   Also known as Sea Goat.   { ¦kap·rə¦kȯr·nəs }

capsule   [AERO ENG]   A small, sealed, pressurized cabin with an internal environment that will support human or animal life during extremely high-altitude flight, space flight, or escape.   { 'kap·səl }

capture   [AERO ENG]   The process in which a missile is taken under control by the guidance system.   [ASTROPHYS]   Of a central force field, as of a planet, to overcome by gravitational force the velocity of a passing body and bring the body under the control of the central force field, in some cases absorbing its mass.   { 'kap·chər }

carbon-detonation supernova model   [ASTRON]   A model for a supernova in a star of 4 to 9 solar masses through the explosive ignition of carbon in a high-density, electron-degenerate core by the formation and propagation of a detonation wave.   { 'kär·bən ¸det·ən'ā·shən ¦sü·pər'nō·və ¸mäd·əl }

carbon sequence   [ASTRON]   Wolf-Rayet stars in which carbon emission bands dominate the spectrum.   { 'kär·bən ¸sē·kwəns }

carbon star   [ASTRON]   Any of a class of stars with an apparently high abundance ratio of carbon to hydrogen; a majority of these are low-temperature red giants of the C class.   { 'kär·bən ¸stär }

Carina   [ASTRON]   A constellation, right ascension 9 hours, declination 60°S.   Abbreviated Car.   Also known as Keel.   { kə'rī·nə }

α **Carinae**   See Canopus.   { ¦al·fə kə'rī¸nē }

Carina Nebula   [ASTRON]   A gaseous nebula near the star η Carinae in the Milky Way.   { kə'ri·nə 'neb·yə·lə }

Carme   [ASTRON]   A small satellite of Jupiter with a diameter of about 19 miles (31 kilometers), orbiting with retrograde motion at a mean distance of 1.4×10^7 miles (2.3×10^7 kilometers).   Also known as Jupiter XI.   { 'kär·mā }

carrier rocket   [AERO ENG]   A rocket vehicle used to carry something, as the carrier rocket of the first artificial earth satellite.   { 'kar·ē·ər ¸räk·ət }

Carrington rotation number   [ASTRON]   A method of numbering rotations of the sun based on a mean rotation period of sunspots of 27.2753 days, and starting with rotation number 1 on November 9, 1853.   { 'kar·iŋ·tən rō'tā·shən ¸nəm·bər }

cartographic satellite   [AERO ENG]   An applications satellite that is used to prepare maps of the earth's surface and of the culture on it.   { ¦kärd·ə¦graf·ik 'sad·əl¸īt }

Cartwheel   [ASTRON]   A ring galaxy found in the southern hemisphere.   { 'kärt¸wēl }

Cas   See Cassiopeia.

Cassini's division   [ASTRON]   The gap, 2500 miles (4000 kilometers) wide, that separates ring A from ring B of the planet Saturn.   { kə'sē·nēz di'vizh·ən }

Cassiopeia [ASTRON] A constellation with right ascension 1 hour, declination 60°N. Abbreviated Cas. { ‚kas·ē·ə'pē·ə }

Cassiopeia A [ASTRON] One of the strongest discrete radio sources, located in the constellation Cassiopeia, associated with patches of filamentary nebulosity which are probably remnants of a supernova. { ‚kas·ē·ə'pē·ə 'ā }

Castor [ASTRON] A multiple star of spectral classification A0 in the constellation Gemini; the star α Geminorum. { 'kas·tər }

cataclysmic variable [ASTRON] **1.** A star showing a sudden increase in the magnitude of light, followed by a slow fading of light; examples are novae and supernovae. Also known as explosive variable. **2.** In particular, a short-period binary star, one of whose components is a white dwarf star, capable of irregularly timed but recurrent outbursts of brightness by 2 to 10,000. { ¦kad·ə¦kliz·mik 'ver·ē·ə·bəl }

catalog number [ASTRON] The designation of a star composed of the name of a particular star catalog and the number of the star as listed there. { 'kad·əl‚äg ‚nəm·bər }

CD galaxy [ASTRON] A supergiant elliptical galaxy with an extended envelope, the largest known type of galaxy. { ¦sē¦dē 'gal·ik·sē }

celestial body [ASTRON] Any aggregation of matter in space constituting a unit for astronomical study, as the sun, moon, a planet, comet, star, or nebula. Also known as heavenly body. { sə'les·chəl 'bäd·ē }

celestial coordinates [ASTRON] Any set of coordinates, such as zenithal distance, altitude, celestial latitude, celestial longitude, local hour angle, azimuth and declination, used to define a point on the celestial sphere. { sə'les·chəl kō'ord·nəts }

celestial equator [ASTRON] The primary great circle of the celestial sphere in the equatorial system, everywhere 90° from the celestial poles; the intersection of the extended plane of the equator and the celestial sphere. Also known as equinoctial. { sə'les·chəl i'kwād·ər }

celestial equator system of coordinates See equatorial system. { sə'les·chəl i'kwād·ər ¦sis·təm əv kō'ord·nəts }

celestial globe [ASTRON] A small globe representing the celestial sphere, on which the apparent positions of the stars are located. Also known as star globe. { sə'les·chəl 'glōb }

celestial horizon [ASTRON] That great circle of the celestial sphere which is formed by the intersection of the celestial sphere and a plane through the center of the earth and is perpendicular to the zenith-nadir line. Also known as rational horizon. { sə'les·chəl hə'rīz·ən }

celestial latitude [ASTRON] Angular distance north or south of the ecliptic; the arc of a circle of latitude between the ecliptic and a point on the celestial sphere, measured northward or southward from the ecliptic through 90°, and labeled N or S to indicate the direction of measurement. Also known as ecliptic latitude. { sə'les·chəl 'lad·ə‚tüd }

celestial longitude [ASTRON] Angular distance east of the vernal equinox, along the ecliptic; the arc of the ecliptic or the angle at the ecliptic pole between the circle of latitude of the vernal equinox and the circle of latitude of a point on the celestial sphere, measured eastward from the circle of latitude of the vernal equinox, through 360°. Also known as ecliptic longitude. { sə'les·chəl 'län·jə‚tüd }

celestial mechanics [ASTROPHYS] The calculation of motions of celestial bodies under the action of their mutual gravitational attractions. Also known as gravitational astronomy. { sə'les·chəl mə'kan·iks }

celestial meridian [ASTRON] A great circle on the celestial sphere, passing through the two celestial poles and the observer's zenith. { sə'les·chəl mə'rid·ē·ən }

celestial parallel See parallel of declination. { sə'les·chəl 'par·ə‚lel }

celestial pole [ASTRON] Either of the two points of intersection of the celestial sphere and the extended axis of the earth, labeled N or S to indicate the north celestial pole or the south celestial pole. { sə'les·chəl 'pōl }

celestial reference system [ASTRON] A system for specifying the locations and times of astronomical objects and events. { si¦les·chəl 'ref·rəns ‚sis·təm }

21

celestial sphere [ASTRON] An imaginary sphere of indefinitely large radius, which is described about an assumed center, and upon which positions of celestial bodies are projected along radii passing through the bodies. { sə'les·chəl 'sfir }

Cen *See* Centaurus.

Centaurus [ASTRON] A constellation with right ascension 13 hours, declination 50°S. Abbreviated Cen. { sen'tȯr·əs }

Centaurus A [ASTRON] A strong, discrete radio source in the constellation Centaurus, associated with the peculiar galaxy NGC 5128. { sen'tȯr·əs 'ā }

Centaurus cluster [ASTRON] A large cluster of galaxies that shows a composite structure, with a concentration of galaxies having recession velocities of about 3000 kilometers (1900 miles) per second, and a weaker concentration having recession velocities of about 4500 kilometers (2800 miles) per second. { sen'tȯr·əs ,kləs·tər }

Centaurus X-3 [ASTROPHYS] A source of x-rays that pulses with a period of 4.8 seconds and is eclipsed every 2.1 days; believed to be a binary star, one of whose members is a rotating neutron star. Abbreviated Cen X-3. { sen'tȯr·əs ,eks 'thrē }

central condensation [ASTRON] The bright, central portion of the coma of a comet, containing one or more nuclei. { 'sen·trəl ,kän·dən'sā·shən }

central control [AERO ENG] The place, facility, or activity at which the whole action incident to a test launch and flight is coordinated and controlled, from the make-ready at the launch site and on the range, to the end of the rocket flight down-range. { 'sen·trəl kən'trōl }

central eclipse [ASTRON] An eclipse in which the eclipsing body passes centrally (midpoints in line) over the body eclipsed. { 'sen·trəl i'klips }

central meridian [ASTRON] The meridian of a planet that crosses the center of the visible face of the planet at a given instant. { 'sen·trəl mə'rid·ē·ən }

central-meridian transit [ASTRON] The passage of an object on the surface of a planet across the central meridian. { 'sen·trəl mə'rid·ē·ən 'tranz·ət }

central peak [ASTRON] A mountain located at the center of the floor of a lunar crater. { 'sen·trəl 'pēk }

Cen X-3 *See* Centaurus X-3.

Cep *See* Cepheus.

Cepheid [ASTRON] One of a subgroup of periodic variable stars whose brightness does not remain constant with time and whose period of variation is a function of intrinsic mean brightness. { 'sē·fē·əd }

Cepheus [ASTRON] A constellation with right ascension 22 hours, declination 70°N. Abbreviated Cep. { 'sē·fē·əs }

Ceres [ASTRON] The largest asteroid, with a diameter of about 960 kilometers, mean distance from the sun of 2.766 astronomical units, and C-type surface composition. { 'sir,ēz }

cesium-ion engine [AERO ENG] An ion engine that uses a stream of cesium ions to produce a thrust for space travel. { 'sē·zē·əm ,ī·ən 'en·jən }

Cet *See* Cetus.

Cetus [ASTRON] A constellation with right ascension 2 hours, declination 10°S. Abbreviated Cet. Also known as Whale. { 'sēd·əs }

Cha *See* Chamaeleon.

Chamaeleon [ASTRON] A constellation, right ascension 11 hours, declination 80°S. Abbreviated Cha. Also spelled Chameleon. { kə'mēl·yən }

chamber pressure [AERO ENG] The pressure of gases within the combustion chamber of a rocket engine. { 'chām·bər ,presh·ər }

chamber volume [AERO ENG] The volume of the rocket combustion chamber, including the convergent portion of the nozzle up to the throat. Also known as chamber capacity. { 'chām·bər ,väl·yəm }

Chameleon *See* Chamaeleon. { kə'mēl·yən }

Chandrasekhar limit [ASTROPHYS] A limiting mass of about 1.44 solar masses above which a white dwarf cannot exist in a stable configuration. { ,chən·drə'shā,kär ,lim·ət }

Chandrasekhar-Schönberg limit [ASTROPHYS] A mass limit for the isothermal, helium core of a main-sequence star above which the star must rapidly increase in radius and evolve away from the main sequence.   Also known as Schönberg-Chandrasekhar limit.   { ˌchən·drə'shā,kär 'shərn,bərg ˌlim·ət }

characteristic chamber length [AERO ENG] The length of a straight, cylindrical tube having the same volume as that of the chamber of a rocket engine if the chamber had no converging section.   { ˌkar·ik·tə'ris·tik 'chäm·bər ˌleŋkth }

characteristic exhaust velocity [AERO ENG] Of a rocket engine, a descriptive parameter, related to effective exhaust velocity and thrust coefficient.   Also known as characteristic velocity.   { ˌkar·ik·tə'ris·tik ig'zòst və'läs·əd·ē }

Charles' Wain See Big Dipper.   { 'chärlz 'wān }

Charon [ASTRON] The only known satellite of Pluto, with an orbital period of 6.387 days, distance from Pluto of approximately 19,600 kilometers, and diameter of approximately 1250 kilometers.   { 'ka·rən }

check flight [AERO ENG] **1.** A flight made to check or test the performance of an aircraft, rocket, or spacecraft, or a piece of its equipment, or to obtain measurements or other data on performance.   **2.** A familiarization flight in an aircraft, or a flight in which the pilot or the aircrew are tested for proficiency.   { 'chek ˌflīt }

chemical pressurization [AERO ENG] The pressurization of propellant tanks in a rocket by means of high-pressure gases developed by the combustion of a fuel and oxidizer or by the decomposition of a substance.   { 'kem·i·kəl ˌpresh·ə·rə'zā·shən }

cherry picker [AERO ENG] A crane used to remove the aerospace capsule containing astronauts from the top of the rocket in the event of a malfunction.   { 'cher·ē ˌpik·ər }

Chiron [ASTRON] An object circling the sun in an eccentric orbit which takes it from inside the orbit of Saturn out to near the orbit of Uranus, and which has a period of 50.7 years.   { 'kī,rän }

Chisel See Caelum.   { 'chiz·əl }

Christmas tree model [ASTRON] An explanation of superluminal motion wherein randomly flashing lights represent the superluminal features.   { 'kris·məs ˌtrē ˌmäd·əl }

chromosphere [ASTRON] A transparent, tenuous layer of gas that rests on the photosphere in the atmosphere of the sun.   { 'krō·mə,sfir }

chromospheric network [ASTRON] A large-scale cellular pattern into which the motion of gas in the chromosphere is ordered by magnetic folds, and which is visible in spectroheliograms taken at the Hα (hydrogen alpha) line at a wavelength of about 656 nanometers and in other spectral regions.   { ˌkrō·mə,sfir·ik 'net,wərk }

CH star [ASTRON] A type of metal-poor carbon star that shows especially strong CH, CN, and C_2 bands in its spectra as well as enhanced bands due to the s-process elements; found in the halo of the Milky Way Galaxy.   { ˌsē'äch ,stär }

chugging [AERO ENG] Also known as bumping; chuffing.   **1.** A form of combustion instability in a rocket engine, characterized by a pulsing operation at a fairly low frequency, sometimes defined as occurring between particular frequency limits.   **2.** The noise that is made in this kind of combustion.   { 'chəg·iŋ }

Cir See Circinus.

Circinus [ASTRON] A constellation, right ascension 15 hours, declination 60°S.   Abbreviated Cir.   Also known as Compasses.   { 'sərs·ən·əs }

circle of declination See hour circle.   { 'sər·kəl əv ,dek·lə'nā·shən }

circle of equal declination See parallel of declination.   { 'sər·kəl əv 'ē·kwəl ,dek·lə'nā·shən }

circle of latitude [ASTRON] A great circle of the celestial sphere passing through the ecliptic poles, and hence perpendicular to the plane of the ecliptic.   Also known as parallel of latitude.   { 'sər·kəl əv 'lad·ə,tüd }

circle of longitude [ASTRON] A circle of the celestial sphere, parallel to the ecliptic.   { 'sər·kəl əv 'län·jə,tüd }

circle of perpetual apparition [ASTRON] That circle of the celestial sphere, centered on the polar axis and having a polar distance from the elevated pole approximately equal to the latitude of the observer, within which celestial bodies do not set.   { 'sər·kəl əv pər'pech·ə·wəl ap·ə'rish·ən }

circle of perpetual occultation |ASTRON| That circle of the celestial sphere, centered on the polar axis and having a polar distance from the depressed pole approximately equal to the latitude of the observer, within which celestial bodies do not rise. { 'sər·kəl əv pər'pech·ə·wəl ,äk·əl'tā·shən }

circle of right ascension See hour circle. { 'sər·kəl əv 'rīt ə'sen·shən }

circular orbit |ASTRON| An orbit comprising a complete constant-altitude revolution around the earth. { 'sər·kyə·lər 'ȯr·bət }

circulating current |ASTRON| A current that circulated in an abnormal direction in the atmosphere of the southern hemisphere of Jupiter between 1919 and 1934; its presence was indicated by the behavior of dark spots in the region. { 'sər·kyə,lād·iŋ 'kər·ənt }

circumlunar |ASTRON| Around the moon; generally applied to trajectories. { ¦sər·kəm'lü·nər }

circummeridian altitude See exmeridian altitude. { ¦sər·kəm·mə'rid·ē·ən 'al·tə,tüd }

circumpolar |ASTRON| Revolving about the elevated pole without setting. { ¦sər·kəm'pō·lər }

circumpolar star |ASTRON| A star with its polar distance approximately equal to or less than the latitude of the observer. { ¦sər·kəm'pō·lər 'stär }

circumstellar disk |ASTRON| A flattened cloud of gas or small particles that undergoes approximately circular motion about a star, and in which the material velocity is determined primarily by the balance of gravity and centrifugal force. { ,sər·kəm¦stel·ər 'disk }

cislunar |ASTRON| Of or pertaining to phenomena, projects, or activity in the space between the earth and moon, or between the earth and the moon's orbit. { ¦sis'lü·nər }

civil day |ASTRON| A mean solar day beginning at midnight instead of at noon; may be based on either apparent solar time or mean solar time. { 'siv·əl 'dā }

civil time |ASTRON| Solar time in a day (civil day) that begins at midnight; may be either apparent solar time or mean solar time. { ¦si·vəl ¦tīm }

civil twilight |ASTRON| The interval of incomplete darkness between sunrise (or sunset) and the time when the center of the sun's disk is 6° below the horizon. { ¦si·vəl ¦twī,līt }

civil year See calendar year. { ¦si·vəl ¦yir }

classical Kuiper Belt object |ASTRON| A member of the Kuiper belt that has a near-circular orbit, almost in the ecliptic plane, and has a period of revolution which is outside the resonances with Neptune's period. { ¦klas·i·kəl 'kī·pər ,belt ,äb·jekt }

classical T Tauri star |ASTRON| A T Tauri star that exhibits strong emission lines in its optical spectrum, emits a strong stellar wind, and accretes material from a circumstellar disk. { ¦klas·ə·kəl ¦tē ¦tȯr·ē 'stär }

Clock See Horologium. { kläk }

clock star |ASTRON| Any star that is used to measure time; always a bright star, whose right ascension is well known. { 'kläk ,stär }

closed ecological system |AERO ENG| A system used in spacecraft that provides for the maintenance of life in an isolated living chamber through complete reutilization of the material available, in particular, by means of a cycle wherein exhaled carbon dioxide, urine, and other waste matter are converted chemically or by photosynthesis into oxygen, water, and food. { ¦klōzd ek·ə'läj·ə·kəl ,sis·təm }

closed universe |ASTRON| A cosmological model in which the volume of the universe is finite and in which the expansion of the universe will slow to a halt billions of years in the future, and the universe will then contract, becoming progressively denser, until it ends in a fireball similar to the big bang. { ¦klōzd 'yü·nə,vərs }

closest approach |ASTRON| **1.** The event that occurs when two planets or other celestial bodies are nearest to each other as they orbit about the sun or other primary. **2.** The place or time of such an event. { 'klō·səst ə'prōch }

closure parameter |ASTRON| The ratio of the actual mean mass density of the observable universe to the critical density for a Friedmann universe. { 'klō·zhər pə,ram·əd·ər }

cluster *See* star cluster.   { 'kləs·tər }

cluster cepheids *See* RR Lyrae stars.   { 'kləs·tər 'sef·ē·ədz }

cluster variables *See* RR Lyrae stars.   { ¦kləs·tər ¦ver·ē·ə·bəlz }

CMa *See* Canis Major.

CMi *See* Canis Minor.

CM Tauri   [ASTRON] A supernova observed by the Chinese and Japanese in 1054; remnants are still seen as the Crab Nebula.   { ¦sē¦em 'taúr·ē }

Coalsack   [ASTRON] An area in one of the brighter regions of the Southern Milky Way which to the naked eye appears entirely devoid of stars and hence dark with respect to the surrounding Milky Way region.   { 'kōl,sak }

co-altitude *See* zenith distance.   { kō'al·tə,tüd }

coasting flight   [AERO ENG] The flight of a rocket between burnout or thrust cutoff of one stage and ignition of another, or between burnout and summit altitude or maximum horizontal range.   { 'kō·stiŋ ,flīt }

cockpit   [AERO ENG] A space in an aircraft or spacecraft where the pilot sits.   { 'käk,pit }

coded-aperture telescope   [ASTRON] A soft gamma-ray telescope that uses a coded mask to image celestial sources; the position of a gamma-ray source is determined by comparing the observed projection pattern of the mask with all possible projection patterns.   { ,kōd·əd ,ap·ə·chər 'tel·ə,skōp }

Col *See* Columba.

cold dark matter   [ASTRON] A hypothetical type of dark matter consisting of particles that would have been in thermal equilibrium while traveling at nonrelativistic velocities in the early universe; possibilities include axions, photinos, gravitinos, heavy magnetic monopoles, and weakly interacting massive particles.   { ¦kōld 'därk ,mad·ər }

cold-flow test   [AERO ENG] A test of a liquid rocket without firing it to check or verify the integrity of a propulsion subsystem, and to provide for the conditioning and flow of propellants (including tank pressurization, propellant loading, and propellant feeding).   { ¦kōld ¦flō ,test }

collapsar   [ASTRON] A black hole that forms during the gravitational collapse of a massive star.   { kə'lap,sär }

collimation error   [ASTRON] The amount by which the angle between the optical axis of a transit telescope and its east-west mechanical axis deviates from 90°.   { ,käl·ə'mā·shən ,er·ər }

collision parameter   [AERO ENG] In orbit computation, the distance between a center of attraction of a central force field and the extension of the velocity vector of a moving object at a great distance from the center.   { kə'lizh·ən pə'ram·əd·ər }

color-color diagram   [ASTRON] A graph whose coordinates are both color indices, showing the distribution of stars or other objects.   { 'kəl·ər 'kəl·ər ,dī·ə,gram }

color equation   [ASTRON] A measure of the color sensitivity and response of a method of observation; photographic, visual, or photoelectric techniques may be employed.   { 'kəl·ər i'kwā·zhən }

color excess   [ASTRON] The difference between the observed color index of a star and the color index corresponding to its spectral type.   { 'kəl·ər 'ek,ses }

color-magnitude diagram   [ASTRON] A graph of the apparent or absolute magnitudes of a group of stars versus their color indices.   { 'kəl·ər ¦mag·nə,tüd ,dī·ə,gram }

Columba   [ASTRON] A constellation, right ascension 6 hours, declination 35°S.   Abbreviated Col.   Also known as Dove.   { kə'ləm·bə }

colure   [ASTRON] A great circle of the celestial sphere through the celestial poles and either the equinoxes or solstices, called respectively the equinoctial colure or the solstitial colure.   { kə'lür }

Com *See* Coma Berenices.

coma   [ASTRON] The gaseous envelope that surrounds the nucleus of a comet.   Also known as head.   { 'kō·mə }

Coma Berenices   [ASTRON] A constellation, right ascension 13 hours, declination 20°N.   Abbreviated Com.   Also known as Berenice's Hair.   { 'kō·mə ,ber·ə'nī·sēz }

Coma cluster   [ASTRON] **1.** A group of over 1000 bright galaxies having a recession velocity of about 4300 miles (6900 kilometers) per second. **2.** An open cluster of about 100 stars at a distance of about 80 parsecs (1.5×10^{15} miles or 2.5×10^{15} kilometers). { 'kō·mə ˌkläs·tər }

Coma supercluster [ASTRON] A supercluster that is centered on the Coma cluster of galaxies and has several extensions, including the Great Wall. { ˈkō·mə 'sü·pər,kläs·tər }

combination spectrum [ASTRON] The composite spectrum characteristic of a symbiotic star. { ˌkäm·bə'nā·shən 'spek·trəm }

combustion chamber [AERO ENG] That part of the rocket engine in which the combustion of propellants takes place at high pressure. Also known as firing chamber. { kəm'bəs·chən ˌchām·bər }

combustion instability [AERO ENG] Unsteadiness or abnormality in the combustion of fuel, as may occur in a rocket engine. { kəm'bəs·chən ˌin·stə'bil·əd·ē }

comes [ASTRON] The smaller star in a binary system. Also known as companion. { 'kō,mēz }

comet [ASTRON] A nebulous celestial body having a fuzzy head surrounding a bright nucleus, one of two major types of bodies moving in closed orbits about the sun; in comparison with the planets, the comets are characterized by their more eccentric orbits and greater range of inclination to the ecliptic. { 'käm·ət }

cometary kilometric radiation [ASTRON] Radio waves detected by space probes encountering Comet Halley, consisting of several different emission patterns ranging from intense, sporadic bursts of broad-band noise to continuously rising and falling tones. { 'käm·ə,ter·ē ˈkil·əˌme·trik ˌrād·ē'ā·shən }

cometary nebula [ASTRON] A fan-shaped reflection nebula that resembles a comet in appearance. { 'käm·ə,ter·ē 'neb·yə·lə }

Comet Biela See Biela Comet. { ˈkäm·ət 'byel·ə }

comet family [ASTRON] Those short-period comets whose aphelia correspond closely to Jupiter's orbit. { 'käm·ət ˌfam·lē }

comet group [ASTRON] A division of comets on the basis of their period; the short-period group (periods of less than 200 years) contains orbits that show a strong preference for the plan of the solar system; the planes of long-period comets are randomly distributed. { 'käm·ət ˌgrüp }

Comet Hale-Bopp [ASTRON] A very large comet which was discovered on July 23, 1995, and reached perihelion on April 1, 1997, when its brightness was magnitude -1. { ˈkäm·ət ˈhāl 'bäp }

Comet Halley See Halley's Comet. { ˈkäm·ət 'ha·lē }

Comet Hyakutake [ASTRON] A comet that passed within about 0.1 astronomical unit of earth in late March 1996. { ˈkäm·ət ˌhyä·kú'tä·ke }

Comet Kohoutek See Kohoutek's Comet. { ˈkäm·ət kə'hō,tek }

Comet Shoemaker-Levy [ASTRON] A comet that was torn apart by an encounter with Jupiter in 1992, and whose pieces collided with Jupiter over a period of a few days around July 22, 1994. { ˈkäm·ət ˈshü,māk·ər 'lē·vē }

command guidance [AERO ENG] The guidance of a missile, rocket, or spacecraft by means of electronic signals sent to receiving devices in the vehicle. { kə'mand ˌgīd·əns }

command module [AERO ENG] The spacecraft module that carries the crew, the main communication and telemetry equipment, and the reentry capsule during cruising flight. { kə'mand ˌmäj·ül }

commensurable motions [ASTRON] Mean motions of the planets, or of the satellites of a planet, which satisfy simple arithmetic relationships. { kə'mens·rə·bəl 'mō·shənz }

commensurate orbits [ASTRON] Orbits of two celestial objects about a common center of gravity so that the period of one is a rational fraction of that of the other. { kə'mench·ə·rət 'or·bəts }

common year [ASTRON] A calendar year of 365 days. { ˈkäm·ən ˈyir }

compact H II region [ASTRON] A region of dense ionized hydrogen in interstellar space, not greater than 1 parsec (1.9×10^{13} miles or 3.1×10^{13} kilometers) in diameter. { 'käm,pakt 'äch 'tü ˌrē·jən }

compact radio source [ASTRON] A source of radio-frequency radiation outside the solar system whose flux at an intermediate radio frequency is dominated by the contribution from a single bright component less than 1 kiloparsec (1.9 × 10^{16} miles or 3.1 × 10^{16} kilometers) in diameter. { 'käm·pakt 'rād·ē·ō ,sȯrs }

companion *See* comes. { kəm'pan·yən }

companion body [AERO ENG] A nose cone, last-stage rocket, or other body that orbits along with an earth satellite or follows a space probe. { kəm'pan·yən ,bäd·ē }

comparison star [ASTRON] A star of known brightness used as a standard for comparison in determining the magnitude of a nearby celestial object. { kəm'par·ə·sən ,stär }

Compasses *See* Circinus. { 'käm·pə·səz }

Compton-Getting effect [ASTROPHYS] The sidereal diurnal variation of the intensity of cosmic rays which would be expected from the rotation of the galaxy if cosmic radiation originated in extragalactic regions and was isotropic in intergalactic space, and if this radiation was unaffected at entry to and passage through the galaxy. { ¦käm·tən 'ged·iŋ i'fekt }

comptonization [ASTRON] The redistribution in the energies of photons in interstellar space that results from their scattering from electrons. { ,käm·tə·nə'zā·shən }

Comstock refraction formula [ASTROPHYS] A formula for the apparent angular displacement of an object outside the earth's atmosphere due to refraction, in terms of the barometric pressure, the temperature of the atmosphere, and the observed zenith distance. { 'käm,stäk ri'frak·shən ,fȯr·mya·lə }

concentric ring structures [ASTRON] A formation on the moon's surface consisting of two craters, one inside the other, with approximately the same center. { kən'sen·trik 'riŋ ,strək·chərz }

configuration [AERO ENG] A particular type of specific aircraft, rocket, or such, which differs from others of the same model by the arrangement of its components or by the addition or omission of auxiliary equipment; for example, long-range configuration or cargo configuration. { kən,fig·yə'rā·shən }

conical point *See* inner Lagrangian point. { 'kän·ə·kəl 'pȯint }

conjunction [ASTRON] **1.** The situation in which two celestial bodies have either the same celestial longitude or the same sidereal hour angle. **2.** The time at which this conjunction takes place. { kən'jəŋk·shən }

consanguineous ring structures [ASTRON] A formation on the moon's surface consisting of two or more craters that are similar in form and very close to each other. { ¦kän·saŋ¦gwin·ē·əs 'riŋ ,strək·chərz }

constant-level balloon [AERO ENG] A balloon designed to float at a constant pressure level. Also known as constant-pressure balloon. { ¦kän·stənt ¦lev·əl bə'lün }

constant of aberration [ASTRON] The maximum aberration of a star observed from the surface of the earth, equal to 20.49 seconds of arc. { 'kän·stənt əv ab·ə'rā·shən }

constellation [ASTRON] **1.** Any one of the star groups interpreted as forming configurations in the sky; examples are Orion and Leo. **2.** Any one of the definite areas of the sky. { ,kän·stə'lā·shən }

constituent day [ASTRON] The duration of one rotation of the earth on its axis with respect to an astre fictif, that is, a fictitious star representing one of the periodic elements in the tidal forces; approximates the length of a lunar or solar day. { kən'stich·ə·wənt 'dā }

constrictor [AERO ENG] The exit portion of the combustion chamber in some designs of ramjets, where there is a narrowing of the tube at the exhaust. { kən'strik·tər }

construction weight [AERO ENG] The weight of a rocket exclusive of propellant, load, and crew if any. Also known as structural weight. { kən'strək·shən ,wāt }

contact binary [ASTRON] A binary system at least one of whose components fills its Roche lobe and in which mass exchange is taking place. { 'kän,takt 'bī,ner·ē }

contiguous arc [ASTRON] A crater arc in which successive craters are in contact. { kən'tig·yə·wəs 'ärk }

contiguous chain [ASTRON] A crater chain in which successive craters are in contact. { kən'tig·yə·wəs 'chān }

contiguous craters [ASTRON] A formation on the moon's surface consisting of two craters in contact; the walls at the point of contact are low, broken, or entirely absent. { kən'tig·yə·wəs 'krād·ərz }

controlled-leakage system [AERO ENG] A system that provides for the maintenance of life in an aircraft or spacecraft cabin by a controlled escape of carbon dioxide and other waste from the cabin, with replenishment provided by stored oxygen and food. { kən'trōld 'lēk·ij ,sis·təm }

control-moment gyro [AERO ENG] An internal momentum storage device that applies torques to the attitude-control system through large rotating gyros. { kən'trōl ,mō·mənt ,jī·rō }

control rocket [AERO ENG] A vernier engine, retrorocket, or other such rocket used to change the attitude of, guide, or make small changes in the speed of a rocket, spacecraft, or the like. { kən'trōl ,räk·ət }

control vane [AERO ENG] A movable vane used for control, especially a movable air vane or jet vane on a rocket used to control flight altitude. { kən'trōl ,vān }

convective zone [ASTROPHYS] A region of instability just below the photosphere of the sun in which part of the heat is carried outward by convective currents. { kən'vek·div ,zōn }

cool star [ASTROPHYS] A low-temperature star, generally visible in the infrared range of the electromagnetic spectrum. { 'kül 'stär }

Copernican principle [ASTRON] The idea that the earth occupies a typical or unexceptional position in the universe. { kə'pər·nə·kən 'prin·sə·pəl }

Copernican system [ASTRON] The system of planetary motions according to Copernicus, who maintained that the earth revolves about an axis once every day and revolves around the sun once every year while the other planets also move in orbits centered near the sun. { kə'pər·nə·kən ,sis·təm }

Cordelia [ASTRON] A satellite of Uranus orbiting at a mean distance of 30,910 miles (49,750 kilometers) with a period of 8 hours 4 minutes, and with a diameter of about 16 miles (26 kilometers); the inner shepherding satellite for the outermost ring of Uranus. { kòr'dēl·yə }

core-halo galaxy [ASTRON] A radio galaxy characterized by a relatively large region of diffuse radio emission surrounding a central region of more intense emission. { 'kòr 'hā·lō ,gal·ik·sē }

corona *See* solar corona. { kə'rō·nə }

Corona Australis [ASTRON] A constellation, right ascension 19 hours, declination 40°S. Abbreviated CrA. Also known as Southern Crown. { kə'rō·nə ò'stral·əs }

Corona Borealis [ASTRON] A constellation, right ascension 16 hours, declination 30°N. Abbreviated CrB. Also known as Northern Crown. { kə'rō·nə bòr·ē'al·əs }

coronagraph [ASTRON] An instrument for photographing the corona and prominences of the sun at times other than at total eclipse. { kə'rō·nə,graf }

coronal green line [ASTRON] The strongest emission line in the visible spectrum of the solar corona, located at a wavelength of 530.3 nanometers and resulting from the emission of iron atoms that have lost 13 of their electrons. { 'kòr·ən·əl grēn ,līn }

coronal hole [ASTRON] A large-scale, apparently open structure in the solar corona, devoid of any soft x-ray emission and surrounded by diverging boundary structures. { kə'rō·nəl ,hōl }

coronal mass ejection [ASTRON] A bubble of gas threaded with magnetic field lines, with dimensions of up to hundreds of thousands of miles, that is ejected from the solar corona over the course of several hours and can disrupt the solar wind, resulting in a geomagnetic storm. { kə,rōn·əl 'mas i,jek·shən }

corotating interaction regions [ASTRON] Regions of enhanced magnetic field bounded by jumps in the solar wind speed that form at distances from the sun greater than 2.5 astronomical units. { 'kō'rō,tād·iŋ ,in·tər'ak·shən ,rē·jənz }

Corvus [ASTRON] A constellation, right ascension 12 hours, declination 20°S. Abbreviated Crv. Also known as Crow. { 'kòr·vəs }

cosmic [ASTRON] Pertaining to the cosmos, the vast extraterrestrial regions of the universe. { 'käz·mik }

cosmic abundance [ASTRON] The amount of a substance believed to be present in the entire universe, relative to other substances. { 'käz·mik ə'bən·dəns }

cosmic background radiation *See* cosmic microwave radiation. { 'käz·mik ¦bak¸graủnd ¸rād·ē¸ā·shən }

cosmic censorship hypothesis [ASTRON] The hypothesis that a system which evolves according to the equations of general relativity from an initial state that does not have singularities or any unusual properties will not develop any space-time singularities that would be visible from large distances. { 'käz·mik 'sen·sər¸ship hī¸päth·ə·səs }

cosmic dust [ASTRON] Fine particles of solid matter forming clouds in interstellar space. { 'käz·mik 'dəst }

cosmic electrodynamics [ASTROPHYS] The science concerned with electromagnetic phenomena in ionized media encountered in interstellar space, in stars, and above the atmosphere. { 'käz·mik i¸lek·trō·də'nam·iks }

cosmic expansion [ASTRON] The recession of all distant galaxies from each other, as manifested in the red shift of their spectral lines. { 'käz·mik ik'span·shən }

cosmic light [ASTRON] The contribution to the brightness of the night sky from all unresolved extragalactic sources. { 'käz·mik 'līt }

cosmic microwave background *See* cosmic microwave radiation. { 'käz·mik 'mī·krō ¸wāv 'bak¸graủnd }

cosmic microwave radiation [ASTRON] A nearly uniform flux of microwave radiation that is believed to permeate all of space and to have originated in the big bang. Also known as cosmic background radiation; cosmic microwave background; microwave background. { 'käz·mik 'mī·krō¸wāv ¸rād·ē'ā·shən }

cosmic radio waves [ASTRON] Radio waves reaching the earth from interstellar or intergalactic sources. { 'käz·mik 'rād·ē·ō ¸wāvz }

cosmic string [ASTRON] A hypothetical relic of the early universe, postulated to have a diameter of the order of 10^{-35} meter and a linear density of the order of 4×10^{22} kilograms per meter, and to be either infinitely long or in the form of a closed curve. { 'käz·mik 'striŋ }

cosmic year [ASTRON] The period of rotation of the Milky Way Galaxy, about 220 million years. { 'käz·mik 'yir }

cosmological [ASTRON] Relating to the overall structure of the universe. { ¦käz·mə¦läj·ə·kəl }

cosmochemistry [ASTROPHYS] The science of the chemistry of the universe, particularly that beyond earth, concerned primarily with inferences on pre-solar-system events, solar nebular processes, and early planetary processes as deduced from minerals in meteorites and from chemical and isotopic compositions of meteorites and their parts. { ¦käz·mō¦kem·ə·strē }

cosmogony [ASTROPHYS] Study of the origin and evolution of specific astronomical systems and of the universe as a whole. { käz'mäg·ə·nē }

cosmological principle [ASTRON] The assumption made in most theories of cosmology that the universe is homogeneous on a large scale. { ¦käz·mə¦läj·ə·kəl 'prin·sə·pəl }

cosmological redshift [ASTRON] The red shift that can be ascribed entirely to the general expansion of space-time initiated by the big bang. { ¦käz·mə¦läj·ə·kəl 'red¸shift }

cosmology [ASTRON] The study of the overall structure of the physical universe. { käz 'mäl·ə·jē }

cosmonaut [AERO ENG] An astronaut in the former Soviet Union. { 'käz·mə¸nȯt }

cotidal hour [ASTRON] The average interval expressed in solar or lunar hours between the moon's passage over the meridian of Greenwich and the following high water at a specified place. { ¸kō'tīd·əl 'aủ·ər }

count [AERO ENG] **1.** To proceed from one point to another in a countdown or plus count, normally by calling a number to signify the point reached. **2.** To proceed in a countdown, for example, T minus 90 and counting. { kaủnt }

countdown [AERO ENG] **1.** The process in the engineering definition, used in leading up to the launch of a large or complicated rocket vehicle, or in leading up to a captive test, a readiness firing, a mock firing, or other firing test. **2.** The act of counting inversely during this process. { 'kaůnt,daůn }

counterglow *See* gegenschein. { 'kaůnt·ər,glō }

counter sun *See* anthelion. { 'kaůnt·ər,sən }

Cowell method [AERO ENG] A method of orbit computation using direct step-by-step integration in rectangular coordinates of the total acceleration of the orbiting body. { 'kaů·əl ,meth·əd }

CrA *See* Corona Australis.

Crab Nebula [ASTRON] A gaseous nebula in the constellation Taurus; an amorphous mass which radiates a continuous spectrum involved in a mesh of filaments that radiate a bright-line spectrum. { 'krab 'neb·yə·lə }

Crab pulsar [ASTRON] A pulsar found in the center of the Crab Nebula with a period of about 0.033 second and that emits radiation at all wavelengths from the radio to the x-ray region. { 'krab 'pəl,sär }

Crane *See* Grus. { krān }

Crater [ASTRON] A constellation, right ascension 11 hours, declination 15°S. Abbreviated Crt. Also known as Cup. { 'krād·ər }

crater arc [ASTRON] A series of lunar craters located along a curved line. { 'krād·ər ,ärk }

crater chain [ASTRON] A series of lunar craters located along a straight line. { 'krād·ər ,chān }

craterlet [ASTRON] A very small lunar crater, with diameter less than about 5 miles (8 kilometers), that still has raised walls. { 'krād·ər·lət }

crater pit [ASTRON] A small lunar crater with no raised walls surrounding it. { 'krād·ər ,pit }

CrB *See* Corona Borealis.

Crepe ring *See* ring C. { 'krāp ,riŋ }

crepuscular rays [ASTRON] Streaks of light radiating from the sun shortly before and after sunset which shine through breaks in the clouds or through irregular spaces along the horizon. { krə'pəs·kyə·lər 'rāz }

crescent phase [ASTRON] A phase of the moon or an inferior planet in which less than half of the visible hemisphere is illuminated. { 'kres·ənt ,fāz }

Cressida [ASTRON] A satellite of Uranus orbiting at a mean distance of 38,380 miles (61,770 kilometers) with a period of 11 hours 9 minutes, and with a diameter of about 41 miles (66 kilometers). { 'kres·əd·ə }

critical density [ASTRON] The mass density above which, it is believed, the expansion of the universe will slow down and reverse. { 'krid·ə·kəl 'den·səd·ē }

critical equatorial velocity [ASTRON] In rotating early-type stars, the velocity at which the centrifugal force at the equator equals the force of gravity there. { 'krid·ə·kəl ,ek·wə'tōr·ē·əl və'läs·əd·ē }

critical velocity [AERO ENG] In rocketry, the speed of sound at the conditions prevailing at the nozzle throat. Also known as throat velocity. { 'krid·ə·kəl və'läs·əd·ē }

Cross *See* Crux. { kròs }

crossed-field accelerator [AERO ENG] A plasma engine for space travel in which plasma serves as a conductor to carry current across a magnetic field, so that a resultant force is exerted on the plasma. { 'kròst ,fēld ik'sel·ə,rād·ər }

Crow *See* Corvus. { krō }

Crt *See* Crater.

Cru *See* Crux.

α Crucis [ASTRON] A double star in the constellation Crux that is 220 light-years from the sun; spectral classification BO.5V. Also known as Acrux. { ¦al·fə 'krü·səs }

β Crucis [ASTRON] A star in the constellation Crux that is 370 light-years from the sun, with magnitude 1.3, spectral classification BO.5IV. Also known as Mimosa. { ¦bā·də 'krü·səs }

Crux [ASTRON] A constellation having four principal bright stars which form the figure of a cross; right ascension 12 hours, declination 60°S. Abbreviated Cru. Also known as Cross; Southern Cross. { krüks }

Crv *See* Corvus.

C-type asteroid [ASTRON] A type of asteroid whose surface is very dark and neutral-colored, and probably is of carbonaceous composition similar to primitive carbonaceous chondritic meteorites. { 'sē ,tīp 'as·tə,róid }

culmination [ASTRON] **1.** The position of a heavenly body when at highest apparent altitude. **2.** For a heavenly body which is continually above the horizon, the position of lowest apparent altitude. { kəl·mə'nā·shən }

Cup *See* Crater. { kəp }

curtate distance [ASTRON] The distance between the earth or the sun and the foot of a perpendicular from a planet or comet to the plane of the earth's orbit. { 'kər,tāt ,dis·təns }

curvature correction [ASTRON] A correction applied to the mean of a series of observations on a star or planet to take account of the divergence of the apparent path of the star or planet from a straight line. { 'kər·və·chər kə'rek·shən }

curve of growth [ASTROPHYS] A graph of the equivalent width of an absorption line versus the number of atoms that produce it. { 'kərv əv 'grōth }

cusp cap [ASTRON] One of the 10 bright areas observed near one of the extremities of the illuminated portion of Venus during the crescent phase. { 'kəsp ,kap }

cutoff [AERO ENG] The shutting off of the propellant flow in a rocket, or the stopping of the combustion of the propellant. { 'kət,óf }

CVn *See* Canes Venatici.

cyanogen absorption [ASTROPHYS] Bands in the absorption spectra of stars at wavelengths near 418 nanometers, caused by atmospheric cyanogen; used as a measure of absolute stellar magnitude. { sī'an·ə·jən əb'sórp·shən }

Cybele [ASTRON] An asteroid with a diameter of about 167 miles (269 kilometers), mean distance from the sun of 3.423 astronomical units, and C-type surface composition. { 'sib·ə·lē }

Cyg *See* Cygnus.

Cygnus [ASTRON] A conspicuous northern summer constellation; the five major stars are arranged in the form of a cross, but the constellation is represented by a swan with spread wings flying southward; right ascension 21 hours, declination 40°N. Abbreviated Cyg. Also known as Northern Cross; Swan. { 'sig·nəs }

Cygnus A [ASTRON] A strong, discrete radio source in the constellation Cygnus, associated with two spiral galaxies in collision. { 'sig·nəs 'ā }

Cygnus loop [ASTRON] A supernova remnant about 17,000 years old, and 30–40 parsecs across and probably about 770 parsecs distant that emits radio waves and x-rays as well as visible light. Also known as Veil Nebula. { 'sig·nəs ,lüp }

Cygnus X-1 [ASTROPHYS] A source of x-rays whose intensity varies in an irregular manner, associated with a weak variable radio source and a ninth-magnitude spectroscopic binary star, designated HDE226868, that consists of a blue supergiant and an invisible companion, which may be a black hole. Abbreviated Cyg X-1. { 'sig·nəs ,eks 'wən }

Cygnus X-3 [ASTROPHYS] A variable source of x-rays, with a period of 4.8 hours, associated with a variable radio source that flared up to enormous levels in September 1972 with no observed increase in x-ray emission. Abbreviated Cyg X-3. { 'sig·nəs ,eks 'thrē }

Cyg X-1 *See* Cygnus X-1.

Cyg X-3 *See* Cygnus X-3.

D

daily aberration *See* diurnal aberration. { ¦dā·lē ab·ə'rā·shən }

dark cloud [ASTRON] A relatively dense, cool cloud of interstellar gas, chiefly molecular, whose dust particles obscure the light of stars behind it. { ¦därk 'klaúd }

dark-eclipsing variables [ASTRON] A binary star system, comprising a bright star and an almost dark companion that revolve about each other. { ¦därk ə,klip·siŋ 'ver·ē·ə·bəlz }

dark matter [ASTRON] Matter that is postulated to exist to explain the rotational motion of the Milky Way Galaxy and other galaxies, to explain the motions of galaxies in clusters, and, in certain cosmological theories, to achieve the critical density of matter in the universe that is just sufficient to close the universe. Also known as missing mass. { 'därk ,mad·ər }

dark nebula [ASTRON] A cloud of solid particles which absorbs or scatters away radiation directed toward an observer and becomes apparent when silhouetted against a bright nebula or rich star field. Also known as absorption nebula. { ¦därk ,neb·yə·lə }

dark of the moon [ASTRON] **1.** The time period of approximately a week at the time of a new moon, when the light of the moon is absent at night. **2.** Any period in which the light of the moon is obscured. { ¦därk əv thə 'mün }

dark star [ASTRON] A star that is not visible but is a part of a binary star system; in particular, a star which causes, in an eclipsing variable, a primary eclipse. { ¦därk ¦stär }

Darwin ellipsoids [ASTRON] Ellipsoidal figures of equilibrium of homogeneous bodies moving about each other in circular orbits, calculated by making certain approximations about their mutual tidal influences. { 'där·wən ə'lip,sóidz }

Davida [ASTRON] An asteroid with a diameter of about 322 kilometers, mean distance from the sun of 3.18 astronomical units, and C-type surface composition. { 'dä·və·də }

dawn [ASTRON] The first appearance of light in the eastern sky before sunrise, or the time of that appearance. Also known as daybreak. { dȯn }

dawn side [ASTRON] That side of a celestial object, such as a planet, which points in the direction of its orbital movement. { dȯn ,sīd }

day [ASTRON] One of various units of time equal to the period of rotation of the earth with respect to one or another direction in space; specific examples are the mean solar day and the sidereal day. { dā }

daybreak *See* dawn. { 'dā,brāk }

dayglow [ASTRON] Airglow of the day sky. { 'dā,glō }

daylight [ASTRON] Light of the day, from sun and sky. { 'dā,līt }

daylight saving meridian [ASTRON] The meridian used for reckoning daylight saving time; generally 15° east of the zone of standard meridian. { ¦dā,līt 'sāv·iŋ mə'rid·ē·ən }

daylight saving noon [ASTRON] Twelve o'clock daylight saving time, or the instant the mean sun is over the upper branch of the daylight saving meridian; during a war, when daylight saving time may be used throughout the year and called war time, the expression war noon applies. Also known as summer noon. { ¦dā,līt 'sāv·iŋ 'nün }

daylight saving time |ASTRON| A variation of zone time, usually 1 hour more advanced than standard time, frequently kept during the summer to make better use of daylight. Also known as summer time. { ¦dā,līt ¦sāv·iŋ ¦tīm }

db galaxy |ASTRON| A dumbbell-shaped radio galaxy, believed to consist of two elliptical nuclei surrounded by a common extended envelope. { ¦dē¦bē 'gal·ik·sē }

December solstice |ASTRON| Winter solstice in the Northern Hemisphere. { di'sem·bər 'säl·stəs }

declination |ASTRON| The angular distance of a celestial object north or south of the celestial equator. { ,dek·lə'nā·shən }

decoupling era |ASTRON| The time about 300,000 years after the big bang when matter and radiation, which had previously been strongly coupled, practically ceased to interact, electrons were able to attach to nuclei and form atoms, and photons could propagate freely. { dē'kəp·liŋ ,ir·ə }

decremental arc |ASTRON| A crater arc in which the diameters of the craters decrease from one end of the arc to the other. { 'dek·rə'ment·əl 'ärk }

decremental chain |ASTRON| A crater chain in which the diameters of the craters decrease from one end of the chain to the other. { 'dek·rə'ment·əl ,chān }

deep space |ASTRON| Space beyond the gravitational influence of the earth. { ¦dēp 'spās }

Deep Space Network |AERO ENG| A spacecraft network operated by NASA which tracks, commands, and receives telemetry for all types of spacecraft sent to explore deep space, the moon, and solar system planets. Abbreviated DSN. { ¦dēp ¦spās 'net,wərk }

deep-space probe |AERO ENG| A spacecraft designed for exploring space beyond the gravitational and magnetic fields of the earth. { ¦dēp ¦spās 'prōb }

deferent |ASTRON| An imaginary circle around the earth, postulated by Ptolemy, in whose circumference a celestial body or its epicycle is supposed to move. { 'def·ə·rənt }

Deimos |ASTRON| A satellite of Mars orbiting at a mean distance of 14,600 miles (23,500 kilometers). { 'dā,mȯs }

de Laval nozzle |AERO ENG| A converging-diverging nozzle used in certain rockets. Also known as Laval nozzle. { də·lä'väl ,näz·əl }

delayed repeater satellite |AERO ENG| Satellite which stores information obtained from a ground terminal at one location, and upon interrogation by a terminal at a different location, transmits the stored message. { di'lād ri¦pēd·ər 'sad·əl,īt }

Delphinus |ASTRON| A northern constellation, right ascension 21 hours, declination 10° north. Also known as Dolphin. { del'fē·nəs }

Delta Cephei |ASTRON| A cepheid variable, from which the name of this type of star is derived; it has a period of 5.3 days. { 'del·tə 'sef·ē,ī }

Delta Scuti stars |ASTRON| A class of pulsating variable stars of spectral type A and with periods of less than 8 hours, relatively small amplitude variations, and masses between 1 and 3 solar masses. { ¦del·tə 'sküd·ē ,stärz }

Dembowska |ASTRON| Possibly the only moderately large asteroid other than Vesta whose surface composition resembles that of achondritic meteorites; has a diameter of about 190 miles (145 kilometers) and a mean distance from the sun of 12.93 astronomical units. { dem'bȯf·skə }

Demon Star See Algol. { 'dē·mən ,stär }

Deneb |ASTRON| A white star of spectral classification A2-Ia in the constellation Cygnus; the star α Cygni. { 'den,eb }

Denebola |ASTRON| A white star of stellar magnitude 2.2, spectral classification A2, in the constellation Leo; the star β Leonis. { də'neb·ə·lə }

density specific impulse |AERO ENG| The product of the specific impulse of a propellant combination and the average specific gravity of the propellants. { 'den·səd·ē spə,sif·ik 'im,pəls }

density-wave theory |ASTROPHYS| A theory explaining the spiral structure of galaxies by a periodic variation in space in the density of matter which rotates with a fixed angular velocity while the angular velocity of the matter itself varies with distance from the galaxy's center. { 'den·səd·ē ,wāv ,thē·ə·rē }

deorbit [AERO ENG] To recover a spacecraft from earth orbit by providing a new orbit which intersects the earth's atmosphere. { dē'òr·bət }

descending node [AERO ENG] That point at which an earth satellite crosses to the south side of the equatorial plane of its primary. Also known as southbound node. [ASTRON] The point at which a planet, planetoid, or comet crosses the ecliptic from north to south. { di'sen·diŋ 'nōd }

descriptive astronomy [ASTRON] Astronomy as presented by graphic and verbal description. { di'skrip·tiv ə'strän·ə·mē }

Desdemona [ASTRON] A satellite of Uranus orbiting at a mean distance of 38,935 miles (62,660 kilometers) with a period of 11 hours 24 minutes, and with a diameter of about 36 miles (58 kilometers). { ,dez·də'mōn·ə }

design gross weight [AERO ENG] The gross weight at takeoff that an aircraft, rocket, or such is expected to have, used in design calculations. { di'zīn ¦grōs 'wāt }

despin [AERO ENG] To stop or reduce the rotation of a spacecraft or one of its components. { ,dē'spin }

Despoina [ASTRON] A satellite of Neptune orbiting at a mean distance of 32,500 miles (52,500 kilometers) with a period of 8.0 hours, and with a diameter of about 110 miles (180 kilometers). { des'pòin·ə }

destruct [AERO ENG] The deliberate action of destroying a rocket vehicle after it has been launched, but before it has completed its course. { di'strəkt }

detached binary [ASTRON] A binary system in which the components do not fill their Roche lobes and have little tidal distortion, and in which significant mass exchange is not taking place. { di'tacht 'bī,ner·ē }

D galaxy [ASTRON] A giant galaxy consisting of an elliptically shaped nucleus surrounded by an unusually large envelope. { 'dē ,gal·ik·sē }

diagram on the plane of the celestial equator See time diagram. { 'dī·ə,gram òn thə 'plān əv thə sə'les·chəl ē'kwäd·ər }

diagram on the plane of the equinoctial See time diagram. { 'dī·ə,gram òn thə 'plān əv thə ē·kwə'näk·chəl }

diamond-ring effect [ASTRON] A phenomenon observed just before and after the central phase of a total solar eclipse, in which the last Baily's bead glows brightly compared with other visible features, and the solar corona forms a band that is visible on the rest of the lunar edge. { ¦dī·mənd 'riŋ i,fekt }

dichotomy [ASTRON] The phase of the moon or an inferior planet at which exactly half of its disk is illuminated and the terminator is a straight line. { dī'käd·ə·mē }

differential correction [ASTRON] A method for finding from the observed residuals minus the computed residuals (O − C) small corrections which, when applied to the orbital elements or constants, will reduce the deviations from the observed motion to a minimum. { ,dif·ə'ren·chəl kə'rek·shən }

diffuse galactic light [ASTRON] Starlight that has been scattered or reflected by interstellar dust near the galactic plane. { də'fyüs gə'lak·tik 'līt }

diffuse nebula [ASTRON] A type of nebula ranging from huge masses presenting relatively high surface brightness down to faint, milky structures that are detectable only with long exposures and special filters; may contain both dust and gas or may be purely gaseous. { də'fyüs 'neb·yə·lə }

diffuse skylight See diffuse sky radiation. { də'fyüs 'skī,līt }

diffuse sky radiation [ASTROPHYS] Solar radiation reaching the earth's surface after having been scattered from the direct solar beam by molecules or suspensoids in the atmosphere. Also known as diffuse skylight; skylight; sky radiation. { də¦fyüs ¦skī ,rād·ē'ā·shən }

Dione [ASTRON] A satellite of Saturn that orbits at a mean distance of 2.35 × 10⁵ miles (3.78 × 10⁵ kilometers) and has a diameter of about 700 miles (1120 kilometers). { 'dī·ə,nē }

dipole anisotropy [ASTRON] A deviation of the equivalent blackbody temperature of the cosmic microwave radiation from its average value which is proportional to the cosine of the angle with some given direction, thus resembling the form of radiation from a dipole antenna.   { 'dī‚pōl ‚an·ə'sä·trə·pē }

directional gyro   [AERO ENG] A flight instrument incorporating a gyro that holds its position in azimuth and thus can be used as a directional reference.   Also known as direction indicator.   { də'rek·shən·əl 'jī·rō }

directional stability   [AERO ENG] The property of an aircraft, rocket, or such, enabling it to restore itself from a yawing or side-slipping condition.   Also known as weathercock stability.   { də'rek·shən·əl stə'bil·əd·ē }

direct motion   [ASTRON] Eastward, or counterclockwise, motion of a planet or other object as seen from the North Pole (motion in the direction of increasing right ascension).   { də¦rekt 'mō·shən }

direct solar radiation   [ASTROPHYS] That portion of the radiant energy received at the actinometer direct from the sun, as distinguished from diffuse sky radiation, effective terrestrial radiation, or radiation from any other source.   { də¦rekt ¦sō·lər rād·ē'ā·shən }

dirty ice   [ASTRON] Interstellar ice particles with particles of graphite or other impurities adsorbed on their surfaces.   { 'dər·dē 'īs }

dirty snowball model   [ASTRON] A model of comet structure in which the nucleus of the comet resembles a large dirty snowball.   { ¦dər·dē 'snō‚bȯl ‚mäd·əl }

disc See disk.   { disk }

discrete radio source   [ASTROPHYS] A source of radio waves coming from a small area of the sky.   { di'skrēt 'rād·ē·ō ‚sȯrs }

disk population   [ASTRON] The older Population I stars such as the sun.   { 'disk ‚päp·yə'lā·shən }

dispersion measure   [ASTRON] A quantity that describes the dispersion of a radio signal, proportional to the product of the density of interstellar electrons and the distance to the source.   { də'spər·zhən ‚mezh·ər }

distance-luminosity relation   [ASTRON] The relation in which the light intensity from a star is inversely proportional to the square of its distance.   { ¦dis·təns lü·mə'näs·əd·ē ri'lā·shən }

distance modulus See modulus of distance.   { 'dis·təns ‚mäj·ə·ləs }

disturbed-sun noise   [ASTROPHYS] Noise at times of sunspot or solar flare activity.   { də¦stərbd 'sən ‚nȯiz }

diurnal aberration   [ASTRON] Aberration caused by the rotation of the earth; its value varies with the latitude of the observer and ranges from zero at the poles to 0.31 second of arc.   Also known as daily aberration.   { dī'ərn·əl ‚ab·ə'rā·shən }

diurnal arc   [ASTRON] That part of a celestial body's diurnal circle which lies above the horizon of the observer.   { dī'ərn·əl 'ärk }

diurnal circle   [ASTRON] The apparent daily path of a celestial body, approximating a parallel of declination.   { dī'ərn·əl 'sər·kəl }

diurnal motion   [ASTRON] The apparent daily motion of a celestial body as observed from a rotating body.   { dī'ərn·əl 'mō·shən }

diurnal parallax See geocentric parallax.   { dī'ərn·əl 'par·ə‚laks }

division circle   [ASTRON] A large circular structure attached to the horizontal axis of a transit circle with accurately calibrated markings; used to determine the inclination of the instrument.   { də'vizh·ən ‚sər·kəl }

docking   [AERO ENG] The mechanical coupling of two or more human-made orbiting objects.   { däk·iŋ }

Dog Star See Sirius.   { 'dȯg ‚stär }

Dolphin See Delphinus.   { 'däl·fən }

dome   [ASTRON] A shallow raised structure on the moon's surface with a smooth convex cross section and a diameter anywhere from a few kilometers up to about 80 kilometers (50 miles).   { dōm }

Dopplergram   [ASTRON] An image of the sun showing line-of-sight (approaching and receding) gas motions, obtained using a spectroheliograph that has been modified to use the Doppler effect.   { 'däp·lər‚gram }

Dorado |ASTRON| A constellation of the southern hemisphere, right ascension 5 hours, declination 65° south. Also known as Swordfish. { də'rä·dō }

double cluster |ASTRON| A pair of globular clusters that are physically close to each other, near the northern boundary of the constellation Perseus. { ¦dəb·əl 'kləs·tər }

double star |ASTRON| A star which appears as a single point of light to the eye but which can be resolved into two points by a telescope. { ¦dəb·əl 'stär }

Dove See Columba. { dəv }

DQ Herculis star See intermediate polar. { ¦dē¦kyü 'hər·kyə·ləs ,stär }

Dra See Draco.

Drac See Draco.

Draco |ASTRON| A long, serpentine constellation that surrounds half of the Little Dipper in the north. Abbreviated Dra; Drac. Also known as Dragon. { 'drā,kō }

draconic month See nodical month. { drə,kän·ik 'mənth }

draconic year See eclipse year. { drə,kän·ik 'yir }

Draconids |ASTRON| Several meteor showers whose radiants lie in the constellation Draco. { drə'kän·ədz }

Draco system |ASTRON| A dwarf elliptical galaxy in the Local Group about 250,000 light-years (1.5×10^{18} miles or 2.4×10^{18} kilometers) distant having a diameter of about 3700 light-years (2.2×10^{16} miles or 3.5×10^{16} kilometers) and consisting chiefly of older stars. { 'drā·kō ,sis·təm }

Dragon See Draco. { 'drag·ən }

Drake equation |ASTRON| An equation which gives the number of advanced technological civilizations curently active in the Galaxy as the product of the rate at which new stars are born in the Galaxy, the probability (actually a product of probabilities) that any one of these stars will possess the necessary conditions for life to originate and to slowly evolve to a technological civilization, and the average longevity of such civilizations. { 'drāk i,kwā·zhən }

Draper catalog |ASTRON| A nine-volume catalog of stars completed in 1924; it gives positions, magnitudes, and spectral classes of 225,300 stars. { 'drā·pər 'kad·əl,äg }

Drift I |ASTRON| A group of stars that tend to move in a stream, traveling in the direction of the constellation Orion; it comprises 60% of the stars whose proper motions are known. { ¦drift ¦wən }

Drift II |ASTRON| A group of stars that tend to move in a stream, traveling in the direction of the constellation Scutum; it comprises 40% of the brighter stars. { ¦drift ¦tü }

drogue |AERO ENG| **1.** A small parachute attached to a body for stabilization and deceleration. Also known as deceleration parachute. **2.** A funnel-shaped device at the end of the hose of a tanker aircraft in flight, to receive the probe of another aircraft that will take on fuel. { drōg }

dry-fuel rocket |AERO ENG| A rocket that uses a mixture of rapidly burning powders; used especially as a booster rocket. { 'drī ¦fyül 'räk·ət }

dry start |AERO ENG| The starting up of a liquid-fuel rocket engine without having previously filled the regenerative cooling tubes. { 'drī 'stärt }

D2 radio source |ASTRON| A radio source consisting of a small, variable nuclear component sometimes coincident with an optical object, and a second, larger component with a much steeper radio spectrum. { ¦dē¦tü 'rād·ē·ō ,sórs }

D-type symbiotic star |ASTRON| A member of a class of symbiotic stars that show infrared emission indicative of astronomical dust, that is, thermal radiation at average temperatures of typically 1000 K; it generally contains Mira variables and has binary periods longer than 10 years. { ¦dē,tīp ,sim·bē,äd·ik 'stär }

Dumbbell Nebula |ASTRON| A planetary nebula of large apparent diameter and low surface brightness in the constellation Vulpecula, about 220 parsecs (4.2×10^{15} miles or 6.8×10^{15} kilometers) away. { 'dəm,bel 'neb·yə·lə }

dusk |ASTRON| That part of either morning or evening twilight between complete darkness and civil twilight. { dəsk }

dusk side |ASTRON| The side of a planet or other celestial body pointing away from its orbital movement direction. { 'dəsk ,sīd }

dust tail [ASTRON] A comet tail that consists of particles, typically 1 micrometer in diameter and primarily silicate in composition, and is usually curved with a length in the range from 10^6 to 10^7 kilometers. { 'dəst ˌtāl }

dwarf Cepheids [ASTRON] A class of pulsating variable stars with periods of less than 6 hours and spectral type A or F; similar to δ Scuti stars but sometimes distinguished from them by the slightly larger amplitudes of their light curves. Also known as Al Velorum stars. { 'dwȯrf 'sef·ē·ədz }

dwarf galaxy [ASTRON] An elliptical galaxy with low mass and low luminosity, having at most a few tens of millions of stars. { 'dwȯrf 'gal·ik·sē }

dwarf novae [ASTRON] A class of irregular variable stars which undergo rapid increases in brightness of several magnitudes at semiperiodic intervals, and then decrease more slowly to the normal minimum; they may be divided into U Geminorum stars and Z Camelopardalis stars. { dwȯrf 'nō,vī }

dwarf spheroidal galaxy [ASTRON] One of the smallest and faintest of the dwarf galaxies, with an effective radius of 200–1000 parsecs and an absolute visual magnitude between −8 and −13. { ¦dwȯrf sfir¦ȯid·əl 'gal·ik·sē }

dwarf star [ASTRON] A star that typically has surface temperature of 5730 K, radius of 428,000 miles (690,000 kilometers), mass of 2×10^{33} grams, and luminosity of 4×10^{33} ergs per second. Also known as main sequence star. { ¦dwȯrf 'stär }

dynamical halo model [ASTRON] A model for the behavior of cosmic rays in the Galaxy in which the cosmic rays are produced in a thin disk near the central plane and then diffuse through the disk and into an outwardly convecting halo to an outer boundary at a distance of perhaps several kiloparsecs from the central plane. { dī'nam·ə·kəl 'hā·lō ˌmäd·əl }

dynamical parallax [ASTRON] A parallax of binary stars that is computed from the sum of the masses of the binary system. { dī¦nam·ə·kəl 'par·əˌlaks }

dynamic load [AERO ENG] With respect to aircraft, rockets, or spacecraft, a load due to an acceleration of craft, as imposed by gusts, by maneuvering, by landing, by firing rockets, and so on. { dī¦nam·ik 'lōd }

dynamic parallax [ASTRON] A value for the parallax of a binary star computed from the observations of the period and angular dimensions of the orbit by assuming a value for the mass of the binary system. Also known as hypothetical parallax. { dī¦nam·ik 'par·əˌlaks }

E

Eagle Nebula [ASTRON] A large emission nebula in the constellation Serpens, about 2500 parsecs away. { 'ē·gəl 'neb·yə·lə }

early-type spiral [ASTRON] A spiral galaxy with a large nuclear bulge and tightly wound arms. { 'ər·lē ¦tīp 'spī·rəl }

early-type star [ASTROPHYS] A star with relatively high surface temperature, in spectral class O or B. { 'ər·lē ¦tīp 'stär }

earth [ASTRON] The third planet in the solar system, lying between Venus and Mars; sometimes capitalized. { ərth }

earthlight [ASTRON] The illumination of the dark part of the moon's disk, produced by sunlight reflected onto the moon from the earth's surface and atmosphere. Also known as earthshine. { 'ərth,līt }

earth orbit [ASTRON] The elliptical motion of the earth about the sun (eccentricity 0.01675, average radius 9.296×10^7 miles or 1.496×10^8 kilometers) in a sidereal year. { ¦ərth 'ȯr·bət }

earth rate [ASTRON] The angular velocity or rate of the earth's rotation. { 'ərth ,rāt }

earth resources technology satellite [AERO ENG] One of a series of satellites designed primarily to measure the natural resources of the earth; functions include mapping, cataloging water resources, surveying crops and forests, tracing sources of water and air pollution, identifying soil and rock formations, and acquiring oceanographic data. Abbreviated ERTS. { ¦ərth ri¦sȯr·səz tek¦näl·ə·je 'sad·əl,īt }

earthrise [ASTRON] The rising of the earth above the horizon of the moon, as viewed from the moon. { 'ərth,rīz }

earth rotation [ASTRON] Motion about the earth's axis that occurs 365.2422 times over a year's period. { ¦ərth rō¦tā·shən }

earth satellite [AERO ENG] An artificial satellite placed into orbit about the earth. [ASTRON] A natural body that revolves about the earth, such as the moon. { 'ərth ,sad·əl,īt }

earthshine See earthlight. { 'ərth,shīn }

earth's way [ASTRON] The angle between the direction of the earth's motion and the apparent direction of a star. { 'ərths ,wā }

east-west effect [ASTRON] The phenomenon due to the fact that a greater number of cosmic-ray particles approach the earth from a westerly direction than from an easterly. { ¦ēst ¦west i,fekt }

eccentric anomaly [ASTRON] For a planet in an elliptical orbit, the eccentric angle corresponding to the planet's location. { ek¦sen·trik ə'näm·ə·lē }

eccentric orbit [ASTRON] An orbit of a celestial body that deviates markedly from a circle. { ek¦sen·trik 'ȯr·bət }

eccentric ring structure [ASTRON] A formation on the moon's surface consisting of two craters, one inside the other, with the inner crater touching the wall of the outer one at one point. { ek¦sen·trik 'riŋ ,strək·chər }

eclipse [ASTRON] **1.** The reduction in visibility or disappearance of a body by passing into the shadow cast by another body. **2.** The apparent cutting off, wholly or partially, of the light from a luminous body by a dark body coming between it and the observer. Also known as astronomical eclipse. { i'klips }

eclipse seasons [ASTRON] The two times when the sun is near enough to one of the nodes of the moon's orbit for eclipses to occur; this positioning occurs at nearly opposite times of the year, and the eclipse seasons vary yearly because of westward regression of the nodes. { i'klips ,sēz·ənz }

eclipse year [ASTRON] The interval between two successive conjunctions of the sun with the same node of the moon's orbit, equal to 346.62 days. Also known as draconic year; nodical year. { i'klips ,yir }

eclipsing binary See eclipsing variable star. { i'klips·iŋ'bī,nər·ē }

eclipsing variable star [ASTRON] A binary star whose orbit is such that every time one star passes between the observer and its companion an eclipse results. Also known as eclipsing binary; photometric binary. { i'klips·iŋ ¦ver·ē·ə·bəl 'stär }

ecliptic [ASTRON] **1.** The apparent annual path of the sun among the stars; the intersection of the plane of the earth's orbit with the celestial sphere. **2.** The plane of the earth's orbit around the sun. { i'klip·tik }

ecliptic coordinate system [ASTRON] A celestial coordinate system in which the ecliptic is taken as the primary and the great circles perpendicular to it are then taken as secondaries. { i'klip·tik kō'ȯrd·ən·ət ,sis·təm }

ecliptic diagram [ASTRON] A diagram of the zodiac indicating positions of certain celestial bodies in the ecliptic region. { i¦klip·tik 'dī·ə,gram }

ecliptic latitude See celestial latitude. { i¦klip·tik 'lad·ə,tüd }

ecliptic limits [ASTRON] The distance of the sun from a node of the moon's orbit such that a solar eclipse cannot occur, or the greatest distance of the moon from a node such that an eclipse of the moon cannot occur. { i¦klip·tik 'lim·əts }

ecliptic longitude See celestial longitude. { i¦klip·tik 'län·jə,tüd }

ecliptic pole [ASTRON] On the celestial sphere, either of two points 90° from the ecliptic. { i¦klip·tik 'pōl }

E corona [ASTRON] The component of the light seen from the solar corona which consists of radiation emitted from the corona itself, as opposed to scattered light. Also known as emission-line corona. { 'ē kə,rō·nə }

Eddington limit [ASTROPHYS] A limit on the radiation emitted by a star above which the star becomes unstable. { 'ed·iŋ·tən ,lim·ət }

Eddington's model [ASTRON] A model of a star in which energy is transported by radiation throughout the star and the ratio of radiation pressure to gas pressure is assumed to be constant. { 'ed·iŋ·tənz ,mäd·əl }

eddy-current damper [AERO ENG] A device used to damp nutation and other unwanted vibration in spacecraft, based on the principle that eddy currents induced in conducting material by motion relative to magnets tend to counteract that motion. { 'ed·ē ,kə·rənt ,dam·pər }

Edgeworth-Kuiper Belt See Kuiper Belt. { ¦ej,wərth 'kī·pər ,belt }

effective radius [ASTRON] The distance from the center of an external galaxy within which half its luminosity is included. { ə¦fek·tiv 'rād·ē·əs }

effective temperature [ASTROPHYS] A measure of the effective temperature of a star, deduced by means of the Stefan-Boltzmann law, from the total energy that is emitted per unit area. { ə¦fek·tiv 'tem·prə·chər }

E galaxy See elliptical galaxy. { 'ē ,gal·ik·sē }

Egeria [ASTRON] An asteroid with a diameter of about 139 miles (224 kilometers), mean distance from the sun of 2.58 astronomical units, and G-type (C-like) surface composition. { e'gir·ē·ə }

Egg Nebula [ASTRON] A reflection nebula consisting of two optical components separated by about 8 arc-seconds, with an infrared source between them. { 'eg ,neb·yə·lə }

egress [ASTRON] The departure of the moon from the shadow of the earth in an eclipse, or of a planet from the disk of the sun, or of a satellite (or its shadow) from the disk of the parent planet. { 'ē,gres }

Elara [ASTRON] A small satellite of Jupiter with a diameter of about 20 miles (32 kilometers), orbiting at a mean distance of 7.29 × 10⁶ miles (11.73 × 10⁶ kilometers). Also known as Jupiter VII. { e'lar·ə }

Electra [ASTRON] A small, irregularly shaped satellite of Saturn that librates about the leading Lagrangian point of Dione's orbit. { i'lek·trə }

electric engine [AERO ENG] A rocket engine in which the propellant is accelerated by some electric device. Also known as electric propulsion system; electric rocket. { i¦lek·trik 'en·jən }

electric propulsion [AERO ENG] A general term encompassing all the various types of propulsion in which the propellant consists of electrically charged particles which are accelerated by electric or magnetic fields, or both. { i¦lek·trik prə'pəl·shən }

electromagnetic propulsion [AERO ENG] Motive power for flight vehicles produced by electromagnetic acceleration of a plasma fluid. { i¦lek·trō·mag'ned·ik prə'pəl·shən }

electropulse engine [AERO ENG] An engine, for propelling a flight vehicle, that is based on the use of spark discharges through which intense electric and magnetic fields are established for periods ranging from microseconds to a few milliseconds; a resulting electromagnetic force drives the plasma along the leads and away from the spark gap. { i'lek·trō,pəls ,en·jən }

electrothermal propulsion [AERO ENG] Propulsion of spacecraft by using an electric arc or other electric heater to bring hydrogen gas or other propellant to the high temperature required for maximum thrust; an arc-jet engine is an example. { i¦lek·trō'thər·məl prə'pəl·shən }

Elektra [ASTRON] An asteroid with a diameter of about 113 miles (182 kilometers), mean distance from the sun of 3.117 astronomical units, and C-type surface composition. { i'lek·trə }

elementary ring structure [ASTRON] A formation on the moon's surface consisting of a simple wall of uniform cross section enclosing a circular area which has the same elevation as the surrounding surface. Also known as simple ring. { ,el·ə'men·trē 'riŋ ,strək·chər }

elements [ASTRON] A set of quantities specifying the orbit of a member of the solar system or of a binary star system, used to calculate the body's position at any time. { 'el·ə·mənts }

elephant trunks [ASTRON] Long, dark regions that encroach into the bright matter in diffuse nebulae, usually bordered by bright rims. { 'el·ə·fənt ,trəŋks }

elevated pole [ASTRON] The celestial pole that appears above the horizon. { 'el·ə,vād·əd 'pōl }

elliptical galaxy [ASTRON] A galaxy whose overall shape ranges from a spheroid to an ellipsoid, without any noticeable structural features. Also known as E galaxy; spheroidal galaxy. { ə'lip·tə·kəl 'gal·ik·sē }

elliptical ring structure [ASTRON] A lunar crater enclosed by a wall that is elliptical in shape. { ə'lip·tə·kəl 'riŋ ,strək·chər }

ellipticity [ASTRON] The difference between the equatorial and polar radii of a planet divided by the mean radius. { ē,lip'tis·əd·ē }

elongation [ASTRON] The difference between the celestial longitude of the moon or a planet, as measured from the earth, and that of the sun. { ē,loŋ'gā·shən }

Emden equation [ASTROPHYS] An equation for stellar structure which arises in a model based on the assumption that the star is a gaseous sphere in adiabatic equilibrium, in which the pressure is proportional to $\rho\gamma$, where ρ is the density and γ is a constant; the equation is $d^2y/dx^2 + (2/x)dy/dx + y^n = 0$, where $n = 1/(\gamma - 1)$, x is proportional to the distance from the center of the sphere, and y is proportional to $\rho^{1/n}$. Also known as Lane-Emden equation. { 'em·dən i,kwā·zhən }

Emden function [ASTROPHYS] A solution of the Emden equation with the boundary conditions $y = 1$ and $dy/dx = 0$ at $x = 0$. Also known as Lane-Emden function. { 'em·dən ,fəŋk·shən }

emersion [ASTRON] The reappearance of a celestial body after an eclipse or occultation. { ē'mer·zhən }

emission-line corona See E corona. { i'mish·ən ¦līn kə'rō·nə }

emission-line galaxy [ASTRON] A galaxy whose spectrum displays narrow, high-excitation emission lines. { i'mish·ən ¦līn 'gal·ik·sē }

emission nebula [ASTRON] A type of bright diffuse nebula whose luminosity results from the excitation and ionization of its gas atoms by ultraviolet radiation from a nearby O- or B-type star. { i'mish·ən 'neb·yə·lə }

Enceladus [ASTRON] A satellite of Saturn orbiting at a mean distance of 153,600 miles (238,000 kilometers). { ‚en·se'lä·dus }

Encke division [ASTRON] A faint line that splits the outer ring of Saturn into two. { 'eŋ·kə də,vizh·ən }

Encke's Comet [ASTRON] A very faint comet with the shortest period of any known comet, 3.3 years. { 'eŋ·kəz 'käm·ət }

energetic solar particles [ASTROPHYS] Electrons and atomic nuclei produced in association with solar flares, with energies mostly in the range 1–100 million electronvolts, but occasionally as high as 15 billion electronvolts. Also known as solar cosmic rays. { ‚en·ər'jed·ik ¦sō·lər 'pärd·i·kəlz }

energy management [AERO ENG] In rocketry, the monitoring of the expenditure of fuel for flight control and navigation. { 'en·ər·jē ‚man·ij·mənt }

engine spray [AERO ENG] That part of a pad deluge that is directed at cooling a rocket's engine during launch. { 'en·jən ‚sprā }

entry corridor [AERO ENG] Depth of the region between two trajectories which define the design limits of a vehicle about to enter a planetary atmosphere, or define the desired landing area (footprint). { 'en·trē ‚kär·ə,dòr }

ephemeris [ASTRON] A periodical publication tabulating the predicted positions of celestial bodies at regular intervals, such as daily, and containing other data of interest to astronomers. Also known as astronomical ephemeris. { ə'fem·ə·rəs }

ephemeris day [ASTRON] A unit of time equal to 86,400 ephemeris seconds (International System of Units). { ə'fem·ə·rəs 'dā }

ephemeris second [ASTRON] The fundamental unit of time of the International System of Units from 1960 until 1968, equal to 1/31556925.9747 of the tropical year defined by the mean motion of the sun in longitude at the epoch 1900 January 0 day 12 hours. { ə'fem·ə·rəs 'sek·ənd }

ephemeris time [ASTRON] The uniform measure of time defined by the laws of dynamics and determined in principle from the orbital motions of the planets, specifically the orbital motion of the earth as represented by Newcomb's Tables of the Sun. Abbreviated E.T. { ə'fem·ə·rəs 'tīm }

Epimetheus [ASTRON] A satellite of Saturn which orbits at a mean distance of 151,000 kilometers (94,000 miles), near Saturn's rings, in nearly the same orbit as Janus, and has an irregular shape with an average diameter of 120 kilometers (75 miles). { ‚ep·ə'mē·thē·əs }

epoch [ASTRON] A particular instant for which certain data are valid; for example, star positions in an astronomical catalog, epoch 1950.0. { 'ep·ək }

equal-areas law [ASTRON] The second of Kepler's laws, which states that the line joining a planet and the sun sweeps over equal areas in equal periods of time. { 'ē·kwəl ¦er·ē·əz ‚lò }

equation of the center [ASTRON] The angle between the actual longitude of the moon and the longitude of an imaginary body that moves with constant angular velocity with the same period as the moon. { i'kwā·zhən əv thə 'sen·tər }

equation of time [ASTRON] The addition of a quantity to mean solar time to obtain apparent solar time; formerly, when apparent solar time was in common use, the opposite convention was used; apparent solar time has annual variation as a result of the sun's inclination in the ecliptic and the eccentricity of the earth's elliptical orbit. { i'kwā·zhən əv 'tīm }

equatorial acceleration [ASTROPHYS] A state in which the equatorial atmosphere of a celestial body has a larger absolute angular velocity than the more poleward portions of the atmosphere; exhibited by the sun, Jupiter, and Saturn. { ‚e·kwə'tòr·ē·əl ak,sel·ə'rā·shən }

equatorial horizontal parallax [ASTRON] The parallax of a member of the solar system measured from positional observations made at the same time at two stations on earth, whose distance apart is the earth's equatorial radius. { ‚e·kwə'tòr·ē·əl ‚här·ə¦zänt·əl 'par·ə,laks }

equatorial orbit [ASTRON] An orbit in the plane of the earth's equator. { ‚e·kwə'tȯr· ē·əl 'ȯr·bət }

equatorial plane [ASTRON] The plane passing through the equator of the earth, or of another celestial body, perpendicular to its axis of rotation and equidistant from its poles. { ‚e·kwə'tȯr·ē·əl 'plān }

equatorial system [ASTRON] A set of celestial coordinates based on the celestial equator as the primary great circle; usually declination and hour angle or sidereal hour angle. Also known as celestial equator system of coordinates; equinoctial system of coordinates. { ‚e·kwə'tȯr·ē·əl 'sis·təm }

equinoctial See celestial equator. { ‚ē·kwə'näk·shəl }

equinoctial colure [ASTRON] The great circle of the celestial sphere through the celestial poles and the equinoxes; the hour circle of the vernal equinox. { ‚ē·kwə'näk·shəl kə'lür }

equinoctial point See equinox. { ‚ē·kwə'näk·shəl 'pȯint }

equinoctial system of coordinates See equatorial system. { ‚ē·kwə'näk·shəl ¦sis·təm əv kō'ȯrd·ən·əts }

equinox [ASTRON] **1.** Either of the two points of intersection of the ecliptic and the celestial equator, occupied by the sun when its declination is 0°. Also known as equinoctial point. **2.** That instant when the sun occupies one of the equinoctial points. { 'ē·kwə‚näks }

Equl See Equuleus.

Equuleus [ASTRON] A northern constellation near Aquarius, right ascension 21 hours, declination 10° north. Abbreviated Equl. Also known as Little Horse. { e'kwül· ē·əs }

Eri See Eridanus.

Erid See Eridanus.

Eridanus [ASTRON] A southern constellation made up of a long, crooked line of stars beginning near Rigel in the foot of Orion, and winding west and south to the first-magnitude star Achernar. Abbreviated Eri; Erid. Also known as River Po. { ‚er· ə'dan·əs }

Eros [ASTRON] The first asteroid to be orbited and landed on by a spacecraft in 2000–2001; the elongated object's maximum diameter is about 19.6 miles (31.6 kilometers); its closest approach to the earth is at about 14×10^6 miles (22.5×10^6 kilometers). { 'e‚räs }

eruptive prominence [ASTRON] A prominence on the sun that is formed from active material above the chromosphere and reaches high altitudes on the sun at great speed. { i'rəp·tiv 'präm·ə·nəns }

eruptive star [ASTRON] A star that has a rapid change in its intensity because of the physical change it undergoes; examples are flare stars, recurrent novae, novae, supernovae, and nebular variables. { ə'rəp·tiv 'stär }

escape orbit [ASTRON] One of various paths that a body or particle escaping from a central force field must follow in order to escape. { ə'skāp ‚ȯr·bət }

escape rocket [AERO ENG] A small rocket engine attached to the leading end of an escape tower, to provide additional thrust to the capsule in an emergency; it helps separate the capsule from the booster vehicle and carries it to an altitude where parachutes can be deployed. { ə'skāp ‚räk·ət }

escape tower [AERO ENG] A trestle tower placed on top of a space capsule, connecting the capsule to the escape rocket on top of the tower; used for emergencies. { ə'skāp ‚taȯr }

escape velocity [ASTRON] The minimum speed away from a parent body that a particle must acquire to escape permanently from the gravitational attraction of the parent. { ə'skāp və‚läs·əd·ē }

estival [ASTRON] Of or pertaining to the summer. Also spelled aestival. { 'es·tə·vəl }

E.T. See ephemeris time.

Eugenia [ASTRON] An asteroid with a diameter of about 133 miles (215 kilometers), mean distance from the sun of 2.72 astronomical units, and F-type (C-like) surface composition; observed to have a small satellite orbiting around it. { yü'jēn·yə }

Eunomia

Eunomia [ASTRON] An asteroid with a diameter of about 161 miles (259 kilometers), mean distance from the sun of 2.64 astronomical units, and S-type surface composition. { yü'nō·mē·ə }

Euphrosyne [ASTRON] An asteroid with a diameter of about 154 miles (248 kilometers), mean distance from the sun of 3.15 astronomical units, and B-type (C-like) surface composition. { yü'fräz·ən·ē }

Eureka [ASTRON] An asteroid orbiting near the following Lagrangian point of the planet Mars, located on Mars's orbit, 60° behind the planet; the first Trojan asteroid of the sun-Mars system to be discovered, in 1990. { yə'rē·kə }

Europa [ASTRON] **1.** A satellite of Jupiter with a mean distance from Jupiter of 4.17 × 10⁵ miles (6.71 × 10⁵ kilometers), orbital period of 3.6 days, and diameter of about 1950 miles (3100 kilometers). Also known as Jupiter II. **2.** An asteroid with a diameter of about 183 miles (295 kilometers), mean distance from the sun of 3.09 astronomical units, and C-type surface composition. { yù'rō·pə }

evection [ASTROPHYS] A perturbation of the moon in its orbit due to the attraction of the sun. { ē'vek·shən }

evening star [ASTRON] A misnomer for a planet that can be seen without a telescope when it sets after the sun. { ¦ev·niŋ 'stär }

evening twilight [ASTRON] The period of time between sunset and darkness. { ¦ev·niŋ 'twī͵līt }

Evershed effect [ASTRON] A displacement of spectral lines of sunspots near the sun's limb, caused by outward motion of gases from the center of the sunspot. { 'ev·ər͵shed i͵fekt }

exhaust stream [AERO ENG] The stream of matter or radiation emitted from the nozzle of a rocket or other reaction engine. { ig'zòst ͵strēm }

exmeridian altitude [ASTRON] An altitude of a celestial body near the celestial meridian of the observer to which a correction is to be applied to determine the meridian altitude. Also known as circummeridian altitude. { eks·mə'rid·ē·ən 'al·tə͵tüd }

exmeridian observation [ASTRON] **1.** Measurement of the altitude of a celestial body near the celestial meridian of the observer, for conversion to a meridian altitude. **2.** The altitude so measured. { eks·mə'rid·ē·ən ͵äb·zər'vā·shən }

exoplanet *See* extrasolar planet. { ͵ek·sō'plan·ət }

exozodiacal dust [ASTRON] Dust that appears to be orbiting within a few astronomical units of a star and may be analogous to the zodiacal dust in the solar system. { ͵ek·sō·zō͵dī·ə·kəl 'dəst }

expandable space structure [AERO ENG] A structure which can be packaged in a small volume for launch and then erected to its full size and shape outside the earth's atmosphere. { ik¦span·də·bəl 'spās ͵strək·chər }

expanding arm [ASTRON] A spiral arm of the Galaxy consisting of neutral hydrogen that lies between 2.5 and 4 kiloparsecs beyond the galactic center and is moving out from it at about 85 miles (135 kilometers) per second. { ik'spand·iŋ 'ärm }

expanding universe [ASTROPHYS] Explanation of the red shift observed in spectral lines from distant galaxies as due to a mutual recession of galaxies away from each other. { ik¦spand·iŋ 'yü·nə·vərs }

expendable vehicle [AERO ENG] A rocket that is used only once to place a payload in orbit. Also known as expendable rocket. { ik'spen·də·bəl 'vē·ə·kəl }

explosive bolt [AERO ENG] A bolt designed to contain a remote-initiated explosive charge which, upon detonation, will shear the bolt or cause it to fail otherwise; applicable to such uses as stage separation of rockets, jettison of expended fuel tanks, and ejection of parachutes. { ik'splō·siv 'bōlt }

explosive decompression [AERO ENG] A sudden loss of pressure in a pressurized cabin, cockpit, or the like, so rapid as to be explosive, as when punctured by gunfire. { ik'splō·siv ͵dē·kəm'presh·ən }

explosive nucleosynthesis [ASTRON] Nucleosynthetic processes that are believed to occur in novae and supernovae, and at the surfaces of neutron stars, such as the *r*-process and the *rp*-process. { ik'splō·siv ͵nü·klē·ō'sin·thə·səs }

explosive variable *See* cataclysmic variable. { ik'splō·siv 'ver·ē·ə·bəl }

extended source [ASTRON] A radio source that has a large angular extent and is strongest at longer wavelengths, as distinguished from a compact source. { ik'stend·əd 'sȯrs }

extensive air shower *See* Auger shower. { ik¦sten·siv 'er ‚shau̇·ər }

extensive shower *See* Auger shower. { ik'sten·siv 'shau̇r }

external galaxy [ASTRON] Any galaxy known to exist, besides the Milky Way. { ek¦stərn·əl 'gal·ik·sē }

extinction [ASTRON] The reduction in the apparent brightness of a celestial object due to absorption and scattering of its light by the atmosphere and by interstellar dust; it is greater at low altitudes. { ek'stiŋk·shən }

extragalactic [ASTRON] Beyond the Milky Way. { ¦ek·strə·gə'lak·tik }

extragalactic background light [ASTRON] The contribution to the brightness of the sky from light coming from outside the Galaxy, chiefly distant galaxies, excluding resolved galaxies. { ¦ek·strə·gə¦lak·tik 'bak‚grau̇nd ‚līt }

extragalactic radio source [ASTROPHYS] A source of radio emission outside the Milky Way. { ¦ek·strə·gə'lak·tik 'rad·ē·ō ‚sȯrs }

extrasolar planet [ASTRON] A planet in orbit about a star other than the sun. Also known as exoplanet. { ¦ek·strə¦sō·lər 'plan·ət }

extraterrestrial intelligence [ASTRON] The potential existence beyond the earth of other advanced civilizations with a technology at least as developed as that on earth. { ¦ek·strə·tə'res·trē·əl in'tel·ə·jəns }

extraterrestrial radiation [ASTROPHYS] Electromagnetic radiation which originates outside the earth or its atmosphere, as in the sun or stars. { ¦ek·strə·tə'res·trē·əl ‚rād·ē'ā·shən }

extravehicular activity [AERO ENG] Activity conducted outside a spacecraft during space flight. { ¦ek·strə·və'hik·yə·lər ak'tiv·əd·ē }

extreme ultraviolet astronomy [ASTRON] Astronomical observations carried out in the region of the electromagnetic spectrum with wavelengths from approximately 10 nanometers to the ionization edge of hydrogen at 91.2 nanometers. { ik¦strēm ‚əl·trə¦vī·lət ə'strän·ə·mē }

extrinsic variable star [ASTRON] A variable star, such as an eclipsing variable, whose variation in apparent brightness is due to some external cause, rather than to actual variaiton in the amount of radiation emitted. { ek¦strinz·ik ‚ver·ē·ə·bəl 'stär }

F

Faber-Jackson relation [ASTRON] A relation between the spectral dispersion caused by the random motions of stars in an elliptical galaxy and the galaxy's intrinsic luminosity. { ¦fāb·ər 'jak·sən ri,lā·shən }

facula [ASTRON] Any of the large patches of bright material forming a veined network in the vicinity of sunspots; faculae appear to be more permanent than sunspots and are probably due to elevated clouds of luminous gas. { 'fak·yə·lə }

failed star See brown dwarf. { ¦fāld 'stär }

fairing [AERO ENG] A structure or surface on an aircraft or rocket that functions to reduce drag, such as the streamlined nose of a satellite-launching rocket. { 'fer·iŋ }

falcate [ASTRON] Crescent-shaped; applied usually to the appearance of the moon, Venus, and Mercury during their crescent phases. { 'fal,kat }

fall [ASTRON] **1.** Of a spacecraft or spatial body, to drop toward a spatial body under the influence of its gravity. **2.** See autumn. { fȯl }

fallaway section [AERO ENG] A section of a rocket vehicle that is cast off from the vehicle during flight, especially such a section that falls back to the earth. { 'fȯl·ə,wā ,sek·shən }

fast nova [ASTRON] A nova whose brightness rises quickly to a maximum, remains near maximum for a short time, and then decreases to the original value in a few years or less. { ¦fast 'nō·və }

fast pulsar [ASTROPHYS] A pulsar with a very short period, of the order of a millisecond. Also known as millisecond pulsar. { 'fast 'pəl,sär }

F corona [ASTRON] The outer portion of the solar corona, consisting of sunlight that has been scattered from interplanetary dust between the sun and the earth. Also known as Fraunhofer corona. { 'ef kə,rō·nə }

fictitious year [ASTRON] The period between successive returns of the fictitious mean sun to a sidereal hour angle of 80° (right ascension 18 hours 40 minutes; about January 1); the length of the fictitious year is the same as that of the tropical year, since both are based upon the position of the sun with respect to the vernal equinox. Also known as Besselian year. { fik'tish·əs 'yir }

field galaxy [ASTRON] An isolated galaxy that does not belong to a cluster. { 'fēld ,gal·ik·sē }

field-line annihilation See field-line reconnection. { 'fēld ,līn ə,nī·ə'lā·shən }

field-line reconnection [ASTRON] A topological rearrangement of the magnetic field lines surrounding an astronomical body, for example, the transfer of lines between open and closed configurations in the terrestrial magnetotail; a possible source of the energy released explosively in solar flares and magnetospheric substorms. Also known as field line annihilation; magnetic merging. { 'fēld ,līn ,rē·kə'nek·shən }

field stars [ASTRON] Background stars when a specific object is being observed. { 'fēld ,stärz }

filament [ASTRON] A prominence, seen as a dark marking on the solar disk. { 'fil·ə·mənt }

filamentary cathode See filament. { ,fil·ə'ment·ə·rē 'kath,ōd }

filament-type cathode See filament. { 'fil·ə·mənt ,tīp 'kath,ōd }

final mass [AERO ENG] The mass of a rocket after its propellants are consumed. { ¦fīn·əl 'mas }

fireball [ASTRON] A bright meteor with luminosity equal to or exceeding that of the brightest planets. { 'fīr,bȯl }

fireball model [ASTROPHYS] A model of gamma-ray bursts according to which a black hole accelerates jets of gas or plasma to relativistic velocities, at about 0.9999 of the speed of light, and the huge amount of kinetic energy in these jets is dissipated in internal collisions within the jets, which produce the gamma-ray emission. { ¦fīr,bȯl ¦mäd·əl }

first lunar meridian [ASTRON] The great circle on the moon which passes through the poles and through the center of the side of the moon that faces the earth. { ¦fərst ¦lün·ər mə'rid·ē·ən }

first motion [AERO ENG] The first indication of motion of a rocket, missile, or test vehicle from its launcher. { 'fərst 'mō·shən }

first point of Aries *See* vernal equinox. { ¦fərst ¦pȯint əv 'er·ēz }

first point of Cancer *See* summer solstice. { ¦fərst ¦pȯint əv 'kan·sər }

first point of Capricorn *See* winter solstice. { ¦fərst ¦pȯint əv 'kap·rə,kȯrn }

first point of Libra *See* autumnal equinox. { ¦fərst ¦pȯint əv 'lē·brə }

first quarter [ASTRON] The phase of the moon when it is near east quadrature, when the western half of it is visible to an observer on the earth. { 'fərst 'kwȯrd·ər }

Fishes *See* Pisces. { 'fish·əz }

fixed star [ASTRON] A misnomer to indicate those stars which kept apparently the same position with respect to other stars, in contrast to the planets which were termed wandering stars. { ¦fikst 'stär }

flame bucket [AERO ENG] A deep, cavelike construction built beneath a rocket launchpad, open at the top to receive the hot gases of the rocket, and open on one or three sides below, with a thick metal fourth side bent toward the open side or sides so as to deflect the exhaust gases. { 'flām ,bək·ət }

flame deflector [AERO ENG] **1.** In a vertical launch, any of variously designed obstructions that intercept the hot gases of the rocket engine so as to deflect them away from the ground or from a structure. **2.** In a captive test, an elbow in the exhaust conduit or flame bucket that deflects the flame into the open. { 'flām di,flek·tər }

Flamsteed's number [ASTRON] A number sometimes used with the possessive form of the Latin name of the constellation to identify a star, for example, 72 Ophiuchi. { 'flam,stēdz ,nəm·bər }

flare stars *See* UV Ceti stars. { 'fler ,stärz }

flash [ASTRON] A thermal instability that occurs in late stages of stellar evolution, according to numerical calculations. { flash }

flash spectrum [ASTRON] The emission spectrum of the sun's chromosphere, observed for a few seconds just before and just after a total solar eclipse. { 'flash ,spek·trəm }

flatness problem [ASTRON] The problem of explaining why, after 10^{10} years of expansion, the density parameter is of the order of 1, whereas the standard big-bang theory suggests that once this parameter deviates even slightly from 1 it very quickly approaches an asymptotic value far away from 1 for open or closed universes. { 'flat·nəs ,präb·ləm }

flicker control [AERO ENG] Control of an aircraft, rocket, or such in which the control surfaces are deflected to their maximum degree with only a slight motion of the controller. { 'flik·ər kən,trōl }

flight [AERO ENG] The movement of an object through the atmosphere or through space, sustained by aerodynamic reaction or other forces. { flīt }

flight characteristic [AERO ENG] A characteristic exhibited by an aircraft, rocket, or the like in flight, such as a tendency to stall or to yaw, or an ability to remain stable at certain speeds. { 'flīt ,kar·ik·tə,ris·tik }

flight path [AERO ENG] The path made or followed in the air or in space by an aircraft, rocket, or such. { 'flīt ,path }

flight science [AERO ENG] The sum total of all knowledge that enables humans to accomplish flight; it is compounded of both science and engineering, and is concerned with airplanes, missiles, and crewed and crewless space vehicles. { 'flīt ,sī·əns }

flight test [AERO ENG] **1.** A test by means of actual or attempted flight to see how an aircraft, spacecraft, space-air vehicle, or missile flies. **2.** A test of a component part of a flying vehicle, or of an object carried in such a vehicle, to determine its suitability or reliability in terms of its intended function by making it endure actual flight. { 'flīt ˌtest }

flocculus [ASTRON] A patch in the sun's surface seen in the light of calcium or hydrogen; the patch may be bright or dark and is usually in the vicinity of sunspots. { 'fläk· yə·ləs }

flux unit [ASTROPHYS] A unit of energy flux density of radio-astronomical sources, equal to 10^{-26} watt per square meter per hertz. Abbreviated fu. { 'fləks ˌyü·nət }

Fly See Musca. { flī }

flyby [AERO ENG] A close approach of a space vehicle to a target planet in which the vehicle does not impact the planet or go into orbit around it. Also known as swing-by. { 'flī,bī }

Flying Fish See Volan. { ¦flī·iŋ ¦fish }

folding fin [AERO ENG] A fin hinged at its base to lie flat, especially a fin on a rocket that lies flat until the rocket is in flight. { 'fōld·iŋ 'fin }

following limb [ASTRON] The half of the limb of a celestial body with an observable disk that appears to follow the body in its apparent motion across the field of view of a fixed telescope. { 'fäl·ə·wiŋ ˌlim }

Fomalhaut See α Piscis Austrini. { 'fō·mə,lòt }

footpad [AERO ENG] A somewhat flat base on the leg of a spacecraft to distribute weight and thereby minimize sinking into a surface. { 'fút,pad }

footpoint [ASTRON] The intersection of tubes of magnetic field lines with the surface of the photosphere. { 'fút,pòint }

Forbush decrease [ASTROPHYS] A sudden decrease in cosmic-ray intensity which occurs a day or two after a solar flare, and at the same time as the commencement of magnetic storms and auroral activity. { 'fòr,bùsh di'krēs }

Fornax A [ASTRON] A peculiar giant elliptical galaxy on the periphery of the Fornax cluster which is a strong double radio source. { 'fòr,naks 'ā }

Fornax cluster [ASTRON] A cluster of galaxies with a few tens of members, about 30 megaparsecs (6×10^{19} miles or 9×10^{19} kilometers) distant. { 'fòr,naks ˌkləs·tər }

Fornax system [ASTRON] A dwarf elliptical galaxy in the Local Group, about 460,000 light-years (2.7×10^{18} miles or 4.4×10^{18} kilometers) distant, having a diameter of about 1600 light-years (9×10^{15} miles or 1.5×10^{16} kilometers) and a mass and luminosity about 7×10^6 that of the sun. { 'fòr,naks ˌsis·təm }

fortnightly nutation [ASTRON] Nutation caused by the change in declination of the moon, having a displacement of up to 0.1 second of arc and a period of 15 days. { ˌfòrt¦nīt·lē nü'tā·shən }

Fortuna [ASTRON] An asteroid with a diameter of about 130 miles (210 kilometers), mean distance from the sun of 2.44 astronomical units, and C-type surface composition. { fòr'tü·nə }

Fraunhofer corona See F corona. { 'fraùn,hōf·ər kə,rō·nə }

free balloon [AERO ENG] A balloon that ascends without a tether, propulsion or guidance; it is made to descend by the release of gas. { ¦frē bə'lün }

French division See Lyot division. { 'french də'vizh·ən }

Friedmann universe [ASTRON] A nonstatic, homogeneous, isotropic model of the universe that displays expansion or contraction and has nonzero matter density. { 'frēd· mən 'yü·nə,vərs }

f spot [ASTRON] One of a pair of sunspots that appears to follow the other across the face of the sun, or whose magnetic polarity is that normally found in such a sunspot during that sunspot cycle and in that hemisphere of the sun. { 'ef ˌspät }

F star [ASTRON] A star whose spectral type is F; surface temperature is 7000 K, and color is yellowish. { 'ef ˌstär }

fu See flux unit.

fuel shutoff [AERO ENG] **1.** The action of shutting off the flow of liquid fuel into a combustion chamber or of stopping the combustion of a solid fuel. **2.** The event or time marking this action. { 'fyül ,shəd,óf }

fuel-weight ratio [AERO ENG] The ratio of the weight of a rocket's fuel to the weight of the unfueled rocket. Also known as fuel structure ratio. { ¦fyül ¦wāt ,rā·shō }

full moon [ASTRON] The moon at opposition, with a phase angle of 0°, when it appears as a round disk to an observer on the earth because the illuminated side is toward the observer. { ¦fúl 'mün }

full pressure suit [AERO ENG] A pressure suit which completely encloses the body and in which a gas pressure sufficiently above ambient pressure for maintenance of function may be sustained. { 'fúl 'presh·ər ,süt }

fundamental star places [ASTRON] The apparent right ascensions and declinations of 1535 standard comparison stars obtained by leading observatories and published annually under the sponsorship of the International Astronomical Union. { ¦fən·də¦ment·əl 'stär ,plās·əz }

funneling [ASTRON] The convergence of the evolutionary paths of stars from different parts of the main sequence into the red giant region on a Hertzsprung-Russell diagram. { 'fən·əl·iŋ }

G

galactic bulge [ASTRON] A spheroidal distribution of stars that is centered on the nucleus of the Milky Way Galaxy and extends to a distance of about 3 kiloparsecs from the center. { gəˈlak·tik ˈbəlj }

galactic center [ASTRON] The gravitational center of the Milky Way Galaxy; the sun and other stars of the Galaxy revolve about this center. { gəˈlak·tik ˈsen·tər }

galactic circle *See* galactic equator. { gəˈlak·tik ˈsər·kəl }

galactic cluster *See* open cluster. { gəˈlak·tik ˈkləs·tər }

galactic concentration [ASTRON] A measure of the increasing density of stars toward the galactic plane, equal to the ratio of the density of stars of a given magnitude at the galactic plane to that at the galactic poles. { gəˈlak·tik ˌkäns·ən·trā·shən }

galactic coordinates *See* galactic system. { gəˈlak·tik kōˈörd·ən·ats }

galactic corona [ASTRON] A low-density gaseous region extending away from the dense gas of the disk of the Milky Way Galaxy into the halo for distances estimated to be at least 3000 parsecs. { gəˈlak·tik kəˈrōn·ə }

galactic disk [ASTRON] The flat distribution of stars and interstellar matter in the spiral arms and plane of the Milky Way Galaxy. { gəˈlak·tik ˈdisk }

galactic equator [ASTRON] A great circle of the celestial sphere, inclined 62° to the celestial equator, coinciding approximately with the center line of the Milky Way, and constituting the primary great circle for the galactic system of coordinates; it is everywhere 90° from the galactic poles. Also known as galactic circle. { gəˈlak·tik iˈkwäd·ər }

galactic halo [ASTRON] The spherical distribution of oldest stars that are centered about the galactic center of the Milky Way Galaxy. { gəˈlak·tik ˈhā·lō }

galactic latitude [ASTRON] Angular distance north or south of the galactic equator; the arc of a great circle through the galactic poles, between the galactic equator and a point on the celestial sphere, measured northward or southward from the galactic equator through 90° and labeled N or S to indicate the direction of measurement. { gəˈlak·tik ˈlad·ə,tüd }

galactic light [ASTRON] The part of the illumination of the night sky that is due to light emitted from stars but diffused through interstellar space. { gəˈlak·tik ˈlīt }

galactic longitude [ASTRON] Angular distance east of sidereal hour angle 94.4° along the galactic equator; the arc of the galactic equator or the angle at the galactic pole between the great circle through the intersection of the galactic equator and the celestial equator in Sagittarius (SHA 94.4°) and a great circle through the galactic poles, measured eastward from the great circle through SHA 94.4° through 360°. { gəˈlak·tik ˈlän·jə,tüd }

galactic nebula [ASTRON] A nebula that is in or near the galactic system known as the Milky Way. { gəˈlak·tik ˈneb·yə·lə }

galactic noise [ASTRON] Radio-frequency noise that originates outside the solar system; it is similar to thermal noise and is strongest in the direction of the Milky Way. { gəˈlak·tik ˈnȯiz }

galactic nova [ASTRON] One of the novae that are concentrated largely in a band 10° on each side of the plane of the galaxy and are most frequent toward the center of the galaxy. { gəˈlak·tik ˈnō·və }

galactic nucleus

galactic nucleus [ASTRON] The center area in the galaxy about which there is a large spherical distribution of stars and from which the spiral arms emanate. { gə'lak·tik 'nü·klē·əs }

galactic plane [ASTRON] The plane that may be drawn through the galactic equator; the plane of the Milky Way Galaxy. { gə'lak·tik 'plān }

galactic pole [ASTRON] On the celestial sphere, either of the two points 90° from the galactic equator. { gə'lak·tik 'pōl }

galactic radiation [ASTROPHYS] Radiation emanating from the Milky Way Galaxy. { gə'lak·tik rād·ē'ā·shən }

galactic rotation [ASTRON] The rotation of the Milky Way about an axis through the center and perpendicular to the plane of the Galaxy; the rotation is apparent from the highly flattened shape and from relative stellar motion. { gə'lak·tik rō'tā·shən }

galactic system [ASTRON] An astronomical coordinate system using latitude measured north and south from the galactic equator, and longitude measured in the sense of increasing right ascension from 0 to 360°. Also known as galactic coordinates. { gə'lak·tik 'sis·təm }

galactic windows [ASTROPHYS] The regions near the equator of the Milky Way where there is low absorption of light by interstellar clouds so that some distant external galaxies may be seen through them. { gə'lak·tik 'win,dōz }

Galatea [ASTRON] A satellite of Neptune orbiting at a mean distance of 38,500 miles (62,000 kilometers) with a period of 10.3 hours, and a diameter of about 90 miles (150 kilometers). { ,gal·ə'tē·ə }

galaxy [ASTRON] A large-scale aggregate of stars, gas, and dust; the aggregate is a separate system of stars covering a mass range from 10^7 to 10^{12} solar masses and ranging in diameter from 1500 to 300,000 light-years. { 'gal·ik·sē }

Galaxy See Milky Way Galaxy. { 'gal·ik·sē }

galaxy cluster [ASTRON] A collection of from two to several hundred galaxies which are much more densely distributed than the average density of galaxies in space. { 'gal·ik·sē ,kləs·tər }

Galilean satellites [ASTRON] The four largest and brightest satellites of Jupiter (Io, Europa, Ganymede, and Callisto). { ,gal·ə'lē·ən 'sad·əl,īts }

gamma-ray astronomy [ASTRON] The study of gamma rays from extraterrestrial sources, especially gamma-ray bursts. { 'gam·ə ,rā ə'strän·ə·mē }

gamma-ray bursts [ASTRON] Intense blasts of soft gamma rays of unknown origin, which range in duration from a tenth of a second to tens of seconds and occur several times a year from sources widely distributed over the sky. { 'gam·ə ,rā ,bərsts }

gamma-ray telescope [ASTRON] Any device for detecting and determining the directions of extraterrestrial gamma rays, using coincidence or anticoincidence circuits with scintillation or semiconductor detectors to obtain directional discrimination. { 'gam·ə ,rā 'tel·ə,skōp }

Ganymede [ASTRON] A satellite of Jupiter orbiting at a mean distance of 664,000 miles (1,071,000 kilometers). Also known as Jupiter III. { 'gan·ə,mēd }

gardening [ASTRON] A phenomenon in which the lunar regolith is constantly churning at a very slow rate because of successive impacts; the result is that bottom material is brought up to the top and surface material is buried. { 'gärd·ən·iŋ }

gas-bounded nebula [ASTRON] An emission nebula whose central star is hot enough, or in which the density of the cloud is small enough, to ionize the entire cloud. { 'gas ¦baún·əd 'neb·yə·lə }

gaseous nebulae [ASTRON] Clouds of gas, such as the Network Nebula in Cygnus, that are members of the Milky Way galactic system and are small compared with its overall dimensions. { ¦gash·əs 'neb·yə·lē }

GAT See Greenwich apparent time.

Gaussian constant [ASTRON] The acceleration caused by the attraction of the sun at the mean distance of the earth from the sun. { ¦gaú·sē·ən 'kän·stənt }

Gaussian year [ASTRON] The period, according to Kepler's laws, of a body of negligible mass traveling in an orbit about the sun whose semimajor axis is 1 astronomical unit; equal to about 365.2569 days. { 'gaús·ē·ən 'yir }

gegenschein [ASTRON] A round or elongated, faint, ill-defined spot of light in the sky at a point 180° from the sun. Also known as counterglow; zodiacal counterglow. { 'gäg·ən,shīn }

Geminga [ASTRON] A relatively nearby neutron star, about 150 parsecs (450 light-years) distant, that emits pulsed x-rays and gamma rays (making it an x-ray and gamma-ray pulsar), steady optical radiation, and possible unconfirmed radio and optical pulsations. The term is derived from Gemini gamma-ray source. { 'jem·iŋ·gə }

Gemini [ASTRON] A northern constellation; right ascension 7 hours, declination 20°N. Also known as Twins. { 'jem·ə,nē }

Geminids [ASTRON] A meteor shower that reaches maximum about December 13. { 'jem·ə·nidz }

general precession [ASTRON] The resultant motion of the components causing precession of the equinoxes westward along the ecliptic at the rate of about 50.3″ per year. { ¦jen·rəl prē'sesh·ən }

geocentric [ASTRON] Relative to the earth as a center, that is, measured from the center of the earth. { ¦jē·ō¦sen·trik }

geocentric coordinates [ASTRON] Coordinates that define the position of a point with respect to the center of the earth; can be either cartesian (x, y, and z) or spherical (latitude, longitude, and radial distance). Also known as geocentric coordinate system; geocentric position. { ¦jē·ō¦sen·trik kō'órd·ən·əts }

geocentric coordinate system See geocentric coordinates. { ¦jē·ō¦sen·trik kō'órd·ən·ət ,sis·təm }

geocentric latitude [ASTRON] The latitude of a celestial body from the center of the earth. { ¦jē·ō¦sen·trik 'lad·ə,tüd }

geocentric longitude [ASTRON] The celestial longitude of the position of a body projected on the celestial sphere when the body is viewed from the center of the earth. { ¦jē·ō¦sen·trik 'län·jə,tüd }

geocentric parallax [ASTRON] The difference in the apparent direction or position of a celestial body, measured in seconds of arc, as determined from the center of the earth and from a point on its surface; this varies with the body's altitude and distance from the earth. Also known as diurnal parallax. { ¦jē·ō¦sen·trik 'par·ə,laks }

geocentric position See geocentric coordinates. { ¦jē·ō¦sen·trik pə'zish·ən }

geocentric zenith [ASTRON] The point where a line from the center of the earth through a point on its surface meets the celestial sphere. { ¦jē·ō¦sen·trik 'zē·nith }

geodetic satellite [AERO ENG] An artificial earth satellite used to obtain data for geodetic triangulation calculations. { ¦jē·ə¦ded·ik 'sad·əl,īt }

geographical position [ASTRON] That point on the earth at which a given celestial body is in the zenith at a specified time. { ¦jē·ə¦graf·ə·kəl pə'zish·ən }

Geographos [ASTRON] An asteroid whose orbit has a semimajor axis of 1.25 astronomical units and an eccentricity of 0.335, giving it a perihelion inside the earth's orbit. { jē'äg·rə,fōs }

geoidal horizon [ASTRON] That circle of the celestial sphere formed by the intersection of the celestial sphere and a plane tangent to the sea-level surface of the earth at the zenith-nadir line. { jē'óid·əl hə'rīz·ən }

geostationary satellite [AERO ENG] A satellite that follows a circular orbit in the plane of the earth's equator from west to east at such a speed as to remain fixed over a given place on the equator at an altitude of 22,280 miles (35,860 kilometers). { ¦jē·ō¦stā·shə,ner·ē 'sad·əl,īt }

geosynchronous orbit [AERO ENG] A satellite orbit that has a period of one sidereal day (23 hours, 56 minutes, 4 seconds). Abbreviated GEO. { ,jē·ō¦siŋ·krə·nəs 'ór·bət }

geosynchronous satellite [AERO ENG] An earth satellite that makes one revolution in one sidereal day (23 hours, 56 minutes, 4 seconds), synchronous with the earth's rotation; the orbit can have arbitrary eccentricity and arbitrary inclination to the earth's equator. Also known as synchronous satellite. { ,jē·ō¦siŋ·krə·nəs 'sad·əl,īt }

GHA See Greenwich hour angle.

Giacobinids [ASTRON] A meteor shower that reaches maximum about October 10, associated with Comet P/Giacobini-Zinner. { jə'kä·bə,nidz }

giant branch [ASTRON] A grouping of stars on the Hertzsprung-Russell diagram that extends upwards and to the right of the main sequence; it represents the first stage of giant-star evolution in which hydrogen fuses to helium in a shell surrounding the core where hydrogen fusion has been exhausted. { ¦jī·ənt 'branch }

giant planets [ASTRON] The planets Jupiter, Saturn, Uranus, and Neptune. { ¦jī·ənt 'plan·əts }

giant star [ASTRON] One of a class of stars that is 20 or 30 or more times larger than the sun and over 100 times more luminous. { ¦jī·ənt 'stär }

gibbous moon [ASTRON] The shape of the moon's visible surface when the sun is illuminating more than half of the side facing the earth. { 'jib·əs 'mün }

gimbaled motor [AERO ENG] A rocket engine mounted on a gimbal. { 'gim·bəld 'mōd·ər }

Giraffe *See* Camelopardalis. { jə'raf }

Gledhill disk [ASTRON] The outer magnetosphere of Jupiter, which forms a disk-shaped region of hot plasma near the plane of Jupiter's magnetic equator. { 'gled,hil ,disk }

Gleissberg cycle [ASTRON] An 80-year cycle in the amplitude of the 11-year sunspot cycle. { 'glīs,bərg ,sī·kəl }

glide path [AERO ENG] **1.** The flight path of an aeronautical vehicle in a glide, seen from the side. Also known as glide trajectory. **2.** The path used by an aircraft or spacecraft in a landing approach procedure. { 'glīd ,path }

glitch [ASTRON] A sudden change in the period of a pulsar, believed to result from a phenomenon analogous to an earthquake that changes the pulsar's moment of inertia. { glich }

globular star cluster [ASTRON] A group of many thousands of stars that are much closer to each other than the stars around the group and that are traveling through space together; a globular cluster has a slightly flattened spheroidal shape. { 'gläb·yə·lər 'stär ,kləs·tər }

globule [ASTRON] A black volume of cosmic dust viewed against the brighter background of bright nebulae. { 'gläb·yəl }

GMT *See* Greenwich mean time.

gnab gib *See* big crunch. { gə'näb ,gib }

gnomic [ASTRON] Pertaining to the gnomon of a sundial. { 'nō·mik }

Gould's belt [ASTRON] A belt of bright stars inclined at 20° to the Milky Way and including most of the bright stars in Orion, Scorpio, Carina, and Centaurus, apparently resulting from a slight tilt of the spiral arm of the Milky Way Galaxy containing the sun, with respect to the galactic plane. { 'gülz ,belt }

gradual phase [ASTROPHYS] The second phase of a solar flare, characterized by emission of relatively low-energy (soft) x-radiation which appears soon after the beginning of the impulsive phase, grows in intensity as the impulsive bursts wane, and lasts up to several hours. Also known as thermal phase. { 'graj·ə·wəl ,fāz }

granulation [ASTRON] The small "rice grain" markings on the sun's photosphere. Also known as photospheric granulation. { ,gran·yə'lā·shən }

granule [ASTRON] A convective cell in the solar photosphere, about 600 miles (1000 kilometers) in diameter. { 'gran·yül }

gravitational astronomy *See* celestial mechanics. { ,grav·ə'tā·shən·əl ə'strän·ə·mē }

gravitational clustering [ASTRON] A theory that attributes the hierarchy structure of the universe to growth of density fluctuations in a statistically uniform and isotropic universe. { ,grav·ə'tā·shən·əl 'kləs·tə,riŋ }

gravitational collapse [ASTRON] The implosion of a star or other astronomical body from an initial size to a size hundreds or thousands of times smaller. { ,grav·ə'tā·shən·əl kə'laps }

gravitational encounter [ASTRON] An approach of two massive bodies in which the directions of motion of both bodies are altered by their mutual gravitational attraction. { ,grav·ə'tā·shən·əl in'kaúnt·ər }

gravitational equilibrium [ASTROPHYS] The condition of a star in which the weight of overlying layers at each point is balanced by the total pressure at that point. { ,grav·ə'tā·shən·əl ,ē·kwə'lib·rē·əm }

gravitational lens   [ASTRON] A massive galaxy or other massive object whose gravitational field focuses light from a distant quasar near or along its line of sight, giving a double or multiple image of the quasar.   { ˌgrav·ə'tā·shən·əl 'lenz }

gravity-gradient attitude control   [AERO ENG] A device that regulates automatically attitude or orientation of an aircraft or spacecraft by responding to changes in gravity acting on the craft.   { ˈgrav·əd·ē ˈgrād·ē·ənt 'ad·ə͵tüd kən͵trōl }

gravity simulation   [AERO ENG] The spinning of part or all of a space vehicle so that the centripetal force on bodies within the vehicle near the outer periphery mimics the force of gravity on objects at the earth's surface.   { 'grav·əd·ē ͵sim·yə͵lā·shən }

grazing-incidence telescope   [ASTRON] An instrument for forming images of celestial x-ray or gamma-ray sources in which the total external reflection of the x-rays or gamma rays from a surface at sufficiently shallow angles of incidence is used to focus them.   { 'grāz·iŋ ˈin·səd·əns 'tel·ə͵skōp }

Great Attractor   [ASTRON] A great supercluster of galaxies and dark matter, approximately 150 × 10⁶ light-years distant, whose existence has been hypothesized to account for the peculiar motions of galaxies, including the Milky Way Galaxy.   { ˈgrāt ə'trak·tər }

great cluster   [ASTRON] A galaxy cluster containing thousands of member galaxies and having a radius of 5 × 10⁶ to 20 × 10⁶ light-years.   { ˈgrāt ˈkləs·tər }

Greater Dog *See* Canis Major.   { 'grād·ər ˈdȯg }

greatest elongation   [ASTRON] The maximum angular distance of a body of the solar system from the sun, as observed from the earth.   { 'grād·əst ͵ē͵loŋ'gā·shən }

Great Nebula of Orion *See* Orion Nebula.   { ˈgrāt 'neb·yə·lə əv ə'rī·ən }

Great Rift   [ASTRON] An apparent break in the Milky Way between Cygnus and Sagittarius caused by a series of large, dark overlapping clouds about 100 parsecs (2×10¹⁵ miles or 3×10¹⁵ kilometers) distant in the equatorial plane of the Galaxy.   { 'grāt 'rift }

Great Wall   [ASTRON] A layer of several thousand galaxies, estimated to extend for about 500 × 10⁶ by 200 × 10⁶ light-years but to be less than 15 × 10⁶ light-years thick, constituting the largest known structure in the universe.   { ˈgrāt 'wȯl }

great year   [ASTRON] The period of one complete cycle of the equinoxes around the ecliptic, about 25,800 years. Also known as platonic year.   { ˈgrāt 'yir }

Greek group   [ASTRON] The group of Trojan planets which lies near the Lagrangian point 60° ahead of Jupiter. Also known as Achilles group.   { 'grēk ˈgrüp }

green flash   [ASTRON] A brilliant green coloration of the upper limb of the sun occasionally observed just as the sun's apparent disk is about to sink below a distant clear horizon. Also known as blue flash; blue-green flame; green segment; green sun.   { 'grēn ˈflash }

green segment *See* green flash.   { 'grēn ˈseg·mənt }

green sun *See* green flash.   { 'grēn ˈsən }

Greenwich apparent noon   [ASTRON] Local apparent noon at the Greenwich meridian; twelve o'clock Greenwich apparent time, or the instant the apparent sun is over the upper branch of the Greenwich meridian.   { 'gren·ich ə'par·ənt ˈnün }

Greenwich apparent time   [ASTRON] Local apparent time at the Greenwich meridian. Abbreviated GAT.   { 'gren·ich ə'par·ənt ˈtīm }

Greenwich civil time *See* Greenwich mean time.   { 'gren·ich 'siv·əl ˈtīm }

Greenwich hour angle   [ASTRON] Angular distance west of the Greenwich celestial meridian; the arc of the celestial equator, or the angle at the celestial pole, between the upper branch of the Greenwich celestial meridian and the hour circle of a point on the celestial sphere, measured westward from the Greenwich celestial meridian through 360°. Abbreviated GHA.   { 'gren·ich 'au̇r ˈaŋ·gəl }

Greenwich interval   [ASTRON] An interval based on the moon's transit of the Greenwich celestial meridian, as distinguished from a local interval based on the moon's transit of the local celestial meridian.   { 'gren·ich 'in·tər·vəl }

Greenwich lunar time   [ASTRON] Local lunar time at the Greenwich meridian; the arc of the celestial equator, or the angle at the celestial pole, between the lower branch of the Greenwich celestial meridian and the hour circle of the moon, measured westward from the lower branch of the Greenwich celestial meridian through 24 hours. { 'gren·ich 'lün·ər ¦tīm }

Greenwich mean noon [ASTRON] Local mean noon at the Greenwich meridian, twelve o'clock Greenwich mean time, or the instant the mean sun is over the upper branch of the Greenwich meridian. { 'gren·ich 'mēn ¦nün }

Greenwich mean time [ASTRON] Mean solar time at the meridian of Greenwich. Abbreviated GMT. Also known as Greenwich civil time; universal time; Z time; zulu time. { 'gren·ich 'mēn ¦tīm }

Greenwich sidereal time [ASTRON] Local sidereal time at the Greenwich meridian. Abbreviated GST. { 'gren·ich sī'dir·ē·əl ¦tīm }

Gregorian calendar [ASTRON] The calendar used for civil purposes throughout the world, replacing the Julian calendar and closely adjusted to the tropical year. { grə'gòr·ē·ən 'kal·ən·dər }

ground start [AERO ENG] A propulsion starting sequence of a rocket or missile that is initiated and carried through to ignition of the main-stage engines on the ground. { 'graund ˌstärt }

ground support equipment [AERO ENG] That equipment on the ground, including all implements, tools, and devices (mobile or fixed), required to inspect, test, adjust, calibrate, appraise, gage, measure, repair, overhaul, assemble, disassemble, transport, safeguard, record, store, or otherwise function in support of a rocket, space vehicle, or the like, either in the research and development phase or in an operational phase, or in support of the guidance system used with the missile, vehicle, or the like. Abbreviated GSE. Also known as ground handling equipment. { 'graund sə,pòrt i¸kwip·məmt }

group [ASTRON] A number of stars moving in the same direction with the same speed. { grüp }

growth factor [AERO ENG] The additional weight of fuel and structural material required by the addition of 1 pound (0.45 kilogram) of payload to the original payload. { 'grōth ˌfak·tər }

Grus [ASTRON] A constellation, right ascension 22 hours, declination 45°S. Also known as Crane. { grüs }

GST See Greenwich sidereal time.

G star [ASTRON] A star of spectral type G; many metallic lines are seen in the spectra, with hydrogen and potassium being strong; G stars are yellow stars, with surface temperatures of 4200–5500 K for giants, 5000–6000 K for dwarfs. { 'jē ˌstär }

guidance system [AERO ENG] The control devices used in guidance of an aircraft or spacecraft. { 'gīd·əns ˌsis·təm }

guillotine factor [ASTROPHYS] A quantity that expresses the sharp reduction in the opacity of a gas which occurs when its temperature becomes sufficiently high to ionize the atoms down to their K shells. { 'gē·ə,tēn ˌfak·tər }

Gum Nebula [ASTRON] A giant nebula about 250 parsecs (5×10^{15} miles or 8×10^{15} kilometers) in diameter, with its near edge about 300 parsecs (6×10^{15} miles or 9×10^{15} kilometers) distant, which is both an old supernova remnant and an H II region. { 'gəm 'neb·yə·lə }

H

Hadar *See* β Centauri. { ha'där }

hadron era [ASTRON] The period in the early universe when the physical forces (gravity, weak, strong, and electromagnetic) diverged from a condition of rough equivalence to increasing disparity, roughly the time between about 10^{-43} and 10^{-4} second after the big bang. { 'had₁rän ₁ir·ə }

Hale cycle [ASTRON] The variation of the sun's magnetic field over a period of approximately 22 years, during which the field reverses and is restored to its original polarity; one such cycle comprises two successive sunspot cycles. { 'hāl ₁sī·kəl }

half-moon [ASTRON] The moon as seen in the first quarter and the last quarter. { 'haf ₁mün }

Halley's Comet [ASTRON] A member of the solar system, with an orbit and a period of about 76 years; its nucleus is about 9 miles (15 kilometers) in diameter; next due to appear in 2061. Also known as Comet Halley. { 'hal·ēz ¦käm·ət }

Hall plasma thruster [AERO ENG] A type of electromagnetic propulsion system in which electrons drawn through an annular discharge chamber feel the effect of a radial magnetic field between the central and outside pole pieces bounding the discharge region, and thereby follow azimuthal cyclotron (Hall-effect) trajectories that increase the probability that they will have ionizing collisions with propellant atoms; the positive ions produced in these collisions are accelerated downstream, thereby producing thrust. { 'hȯl ¦plaz·mə ₁thrəst·ər }

halo [ASTRON] A type of ray system in which many short, filamentary streaks form a complex network of bright matter surrounding the lunar crater. Also known as nimbus. { 'hā·lō }

halo population *See* population II. { 'hā·lō ₁päp·yə₁lā·shən }

Hamal [ASTRON] A second-magnitude star in the constellation Aries; the star α Ari. { hə'mäl }

hang glider [AERO ENG] A flexible, deployable, steerable, kitelike glider from which a harnessed rider hangs during flight. { 'haŋ ₁glīd·ər }

hard landing [AERO ENG] A landing made without deceleration, as by impact on the moon. { 'härd ¦land·iŋ }

Hare *See* Lepus. { her }

harmonic law [ASTRON] The third of Kepler's laws, which states that the squares of the periods of revolution of any two planets are proportional to the cubes of their mean distances from the sun. { här'män·ik 'lȯ }

Harvard-Draper sequence [ASTRON] A system of classification of stellar spectra based on features that are found to vary in a smooth way from one star to another, and on the star's color. Also known as Harvard sequence { ¦här·vərd 'drā·pər ₁sē·kwəns }

Harvard sequence *See* Harvard-Draper sequence.

harvest moon [ASTRON] A full moon that is seen nearest the autumnal equinox. { 'här·vəst ¦mün }

Hayashi track [ASTRON] A vertical track on the Hertzsprung-Russell diagram along which a star of small mass descends during its early stages of formation, when convective heat transport prevails over most of the star. { ha'ya·shē ₁trak }

HdC star [ASTRON] A type of hydrogen-deficient supergiant carbon star that resembles the R Coronae Borealis stars but does not display significant variability. { ˌāch ˌdē'sē ˌstär }

head *See* coma. { hed }

heavenly body *See* celestial body. { 'hev·ən·lē 'bäd·ē }

heavy-metal star [ASTRON] A member of a class of peculiar giants that includes the barium stars and S stars, characterized by unusually strong lines of heavy metals, including barium and zirconium. { 'hev·ē ¦med·əl 'stär }

Hebe [ASTRON] An asteroid with a diameter of about 126 miles (202 kilometers), mean distance from the sun of 2.42 astronomical units, and S-type surface composition. { 'hē·bē }

Hektor [ASTRON] An asteroid, believed to be the largest of the Trojan planets, which circles the sun in the orbit of and approximately 60° ahead of Jupiter; it has an elongated shape, about 186 × 93 miles (300 × 150 kilometers) and D-type surface composition. { 'hek·tər }

heliacal rising [ASTRON] The rising of a celestial body at the same time or just before that of the sun. { hi'lī·ə·kəl 'rīz·iŋ }

heliacal setting [ASTRON] The setting of a celestial body at the same time or just after that of the sun. { hi'lī·ə·kəl 'sed·iŋ }

heliocentric [ASTRON] Relative to the sun as a center. { ¦hē·lē·ō¦sen·trik }

heliocentric coordinates [ASTRON] A coordinate system relative to the sun as a center. { ¦hē·lē·ō¦sen·trik kō'órd·ən·əts }

heliocentric Julian date [ASTRON] The Julian date corrected to the time at which light from the celestial object in question reaches the sun (rather than the earth). Abbreviated HJD. { ˌhē·lē·ō'sen·trik 'jül·yən 'dāt }

heliocentric latitude [ASTRON] Sun-centered coordinate of angular distance perpendicular to the ecliptic plane. { ¦hē·lē·ō¦sen·trik 'lad·ə,tüd }

heliocentric longitude [ASTRON] The angular distance east or west from a given point on the sun's equator. { ¦hē·lē·ō¦sen·trik 'län·jə,tüd }

heliocentric orbit [ASTRON] An orbit relative to the sun as a center. { ¦hē·lē·ō¦sen·trik 'ór·bət }

heliocentric parallax *See* annual parallax. { ¦hē·lē·ō¦sen·trik 'par·ə,laks }

heliographic latitude [ASTRON] On the sun, angular distance north or south of its equator. { ˌhē·lē·ə'graf·ik 'lad·ə,tüd }

heliographic longitude [ASTRON] On the sun, angular distance east or west from given point on the equator of the sun. { ˌhē·lē·ə'graf·ik 'lan·jə,tüd }

heliopause [ASTRON] A shock front about 50 astronomical units from the sun, where the thermal pressure of the interstellar gas which surrounds the solar system overcomes the pressure of the solar wind, which is mostly ram pressure due to its high-velocity flow. { 'hē·lē·ə,póz }

helioseismology [ASTRON] The analysis of wave motions of the solar surface to determine the structure of the sun's interior. { ˌhē·lē·ō,sīz'mäl·ə·jē }

heliosphere [ASTRON] The region surrounding the sun where the solar wind dominates the interstellar medium. Also known as solar cavity. { 'hē·lē·ə,sfir }

helium flash [ASTRON] The onset of runaway helium burning in the degenerate core of a red giant star and the resulting expansion of the core. { 'hē·lē·əm ,flash }

helium stars [ASTRON] The class B stars. { 'hē·lē·əm ¦stärz }

Helix Nebula [ASTRON] A planetary nebula in Aquarius about 140 parsecs (2.7 × 10¹⁵ miles or 4.3 × 10¹⁵ kilometers) distant that has a high helium abundance and the largest known diameter of any planetary nebula. { 'hē,liks 'neb·yə·lə }

Hellas [ASTRON] The largest impact basin on Mars, approximately 1240 miles (2000 kilometers) across and 2.5 miles (4 kilometers) deep, appearing as a bright circular region in earth-based telescopes. { 'hel·əs }

Hellespontus [ASTRON] A surface region of the planet Mars between the regions Hellas and Noachis. { ¦hel·əs¦pän·təs }

Henyey track [ASTRON] An almost horizontal track on the Hertzsprung-Russell diagram that a star of small mass follows in an early stage of evolution after leaving the

Hayashi track and before reaching the main sequence, during which the star is almost wholly in radiative equilibrium. { 'hen·yē ,trak }

Herbig emission star [ASTRON] A relatively massive star in early stages of formation, still surrounded by a nebula which makes it variable in luminosity and renders its spectrum very peculiar. { 'hər·bik i¦mish·ən ,stär }

Herbig-Haro object [ASTRON] A bright patch on the surface of a dark cloud of gas and dust, consisting of light that has been scattered and reflected from a newborn star embedded in the cloud. { ¦hər·big 'hä·rō ,äb·jəkt }

Hercules [ASTRON] A constellation with no stars brighter than third magnitude; right ascension 17 hours, declination 30° north. { 'hər·kyə,lēz }

Hercules cluster [ASTRON] A cluster of about 75 bright galaxies with a recession velocity of 6200 miles (10,000 kilometers) per second. { 'hər·kyə,lēz ,kləs·tər }

Hercules superclusters [ASTRON] A pair of superclusters with recession velocities around 10,000 kilometers (6200 miles) per second, one of which contains the Hercules cluster. { ¦hərk·yə,lēz 'sü·pər,kləs·tərz }

Hercules X-1 [ASTROPHYS] A source of x-rays that pulses with a period of 1.237 seconds, and is eclipsed for 6 of every 42 hours, associated with a variable star, designated HZ Herculis, that also has a period of 42 hours and faint 1.237-second pulsations; believed to be a binary star whose invisible member is a rotating neutron star. Abbreviated Her X-1. { 'hər·kyə,lēz ¦eks 'wən }

Hermes [ASTRON] A very small asteroid which passed within 485,000 miles (780,000 kilometers) of the earth in 1937, the closest known approach of a celestial body other than the moon. { 'hər,mēz }

Hertzsprung gap [ASTRON] A gap on the Hertzsprung-Russell diagram between giant stars of spectral types A0 and G0, caused by the fact that the movement of stars across this region occupies a relatively brief time. { 'hert·sprüŋ ,gap }

Hertzsprung-Russell diagram [ASTRON] A plot showing the relation between the luminosity and surface temperature of stars; other related quantities frequently used in plotting this diagram are the absolute magnitude for luminosity, and spectral type or color index for the surface temperatures. Abbreviated H-R diagram. Also known as Russell diagram. { 'hert·sprüŋ 'rəs·əl 'dī·ə,gram }

Her X-1 See Hercules X-1.

Hesperus [ASTRON] Greek name for the planet Venus as an evening star. { 'hes·prəs }

Hess diagram [ASTRON] A diagram showing the frequencies of occurrence of stars at various positions on the Hertzsprung-Russell diagram. { 'hes ,dī·ə,gram }

Hidalgo [ASTRON] The asteroid with the second largest known mean distance from the sun, about 5.8 astronomical units. { hi'däl·gō }

high-energy astrophysics [ASTROPHYS] A science concerned with studies of acceleration of charged particles to high energies in space, cosmic rays, radio galaxies, pulsars, and quasi-stellar sources. { 'hī ,en·ər·jē as·trə'fiz·iks }

high-mass x-ray binary [ASTRON] A binary system consisting of a massive (greater than 5 solar masses), early-type star and a neutron star or black hole that accretes material through a stellar wind or Roche-lobe overflow, resulting in the emission of hard x-rays. Abbreviated HMXRB. { ¦hī,mas ¦eks,rā'bī,ner·ē }

high-velocity cloud [ASTRON] A rapidly moving interstellar cloud with a radial velocity greater than 12 miles (20 kilometers) per second, consisting primarily of neutral atomic hydrogen, observed in the ultraviolet. { 'hī və'läs·əd·ē 'klaüd }

high-velocity star [ASTRON] A star that moves across the galactic track along which the majority of the stars execute their galactic rotation, thus exhibiting high velocity with respect to the sun and low velocity with respect to the galactic center. { 'hī və¦läs·əd·ē 'stär }

Hilda group [ASTRON] A group of asteroids whose periods of revolution about the sun are approximately two-thirds that of Jupiter, and whose motions are in resonance with Jupiter. { 'hil·də ,grüp }

Hiltner-Hall effect [ASTRON] The polarization of the light received from distant stars; this effect is thought to take place in interstellar space. { ¦hilt·nər 'hȯl i,fekt }

Himalia

Himalia [ASTRON] A small satellite of Jupiter with a diameter of about 35 miles (56 kilometers), orbiting at a mean distance of 7.12×10^6 miles (11.46×10^6 kilometers). Also known as Jupiter VI. { hi'mäl·ē·ə }

Hind's Nebula [ASTRON] A reflection nebula illuminated by the star T Tauri that undergoes marked changes in brightness. { 'hīnz 'neb·yə·lə }

Hirayama family [ASTRON] A clustering of asteroids whose orbits have similar values of semimajor axis, eccentricity, and inclination; over 100 such families have been tabulated. { ‚hi·rä'yä·mä ‚fam·lē }

HJD *See* heliocentric Julian date.

HMXRB *See* high-mass x-ray binary.

Hohmann orbit [AERO ENG] A minimum-energy-transfer orbit. { 'hō·mən ‚ȯr·bət }

Hohmann trajectory [AERO ENG] The minimum-energy trajectory between two planetary orbits, utilizing only two propulsive impulses. { 'hō·mən trə‚jek·tə·rē }

Holmberg radius [ASTRON] The radius of an external galaxy at which the surface brightness is such that the light emitted from one square arc-second equals that from a star of magnitude 26.6. { 'hōm‚bərg ‚rād·ē·əs }

homologous transformation [ASTRON] A mathematical transformation in the study of stellar models. { hə'mäl·ə·gəs ‚tranz·fər'mā·shən }

H I region [ASTRON] A region of interstellar space where neutral hydrogen is present. { ¦āch 'wən ‚rē·jən }

horizon [ASTRON] **1.** The apparent boundary line between the sky and the earth or sea. Also known as apparent horizon. **2.** The distance a light-ray could have traveled since the big-bang explosion at any given epoch in the evolution of the universe. { hə'rīz·ən }

horizon problem [ASTRON] The problem of explaining the observed uniformity of the universe, and in particular of the cosmic background radiation, when, according to the standard big-bang theory, sources of radiation coming from opposite directions in the sky were separated by manyfold the horizon distance at the time of emission, and thus could not possibly have been in physical contact. { hə'rīz·ən ‚präb·ləm }

horizon system of coordinates [ASTRON] A set of celestial coordinates based on the celestial horizon as the primary great circle. { hə'rīz·ən ¦sis·təm əv kō'órd·ən·əts }

horizontal branch [ASTRON] A region in the Hertzsprung-Russell diagram of a typical globular cluster that extends in the blue direction from the giant branch at an absolute bolometric magnitude of 0.3 and consists of stars that are burning helium in their cores and hydrogen in their surrounding envelopes. { ‚här·ə'zänt·əl 'branch }

horizontal parallax [ASTRON] The geocentric parallax of a celestial object when it is rising or setting. { ‚här·ə'zänt·əl 'par·ə‚laks }

Horologium [ASTRON] A constellation with right ascension 3 hours, declination 60° south. Also known as Clock. { ‚hȯr·ə'lō·jē·əm }

Horologium superclusters [ASTRON] Two superclusters in approximately the same direction, with recession velocities around 12,000 kilometers (7500 miles) and 18,000 kilometers (11,200 miles) per second. { ‚hȯr·ə‚lō·gē·əm 'sü·pər‚kləs·tərz }

Horsehead Nebula [ASTRON] A cloud of obscuring particles between the earth and a gaseous emission nebula in the constellation Orion. { 'hȯrs‚hed 'neb·yə·lə }

hot dark matter [ASTRON] A hypothetical type of dark matter consisting of entities that were not in thermal equilibrium in the early universe; possibilities include massive neutrinos and cosmic strings. { ¦hät ¦därk 'mad·ər }

hour angle [ASTRON] Angular distance west of a celestial meridian or hour circle; the arc of the celestial equator, or the angle at the celestial pole, between the upper branch of a celestial meridian or hour circle and the hour circle of a celestial body or the vernal equinox, measured westward through 360°. { 'aȯr ‚aŋ·gəl }

hour-angle difference *See* meridian angle difference. { 'aȯr ‚aŋ·gəl 'dif·rəns }

hour circle [ASTRON] An imaginary great circle passing through the celestial poles on the celestial sphere above which declination is measured. Also known as circle of declination; circle of right ascension. { 'aȯr ‚sər·kəl }

H-R diagram *See* Hertzsprung-Russell diagram. { ¦āch 'är ‚dī·ə‚gram }

H II region [ASTRON] A region of interstellar space occupied by gas that is largely atomic hydrogen and mostly ionized. { ¦āch 'tü ¸rē·jən }

hubble [ASTRON] A unit of astronomical distance equal to 10^9 light-years or 9.4605 × 10^{24} meters. { 'həb·əl }

Hubble constant [ASTROPHYS] The rate at which the velocity of recession of the galaxies increases with distance; the value is about 70 kilometers per second per megaparsec (or 2.3 × 10^{-18} s^{-1}) with a relative uncertainty of about ± 10%. { 'həb·əl ¸kän·stənt }

Hubble effect *See* redshift. { 'həb·əl i¸fekt }

Hubble flow [ASTRON] The mutual recession of celestial objects from each other by virtue of the cosmological expansion of the universe. { 'həb·əl ¸flō }

Hubble law [ASTRON] The principle that the distance of external galaxies from the earth is proportional to their redshift. { 'həb·əl ¸lo }

Hubble's Variable Nebula [ASTRON] A variable-brightness nebula associated with variable stars and fan-shaped in appearance. { 'həb·əlz ¦ver·ē·ə·bəl 'neb·yə·lə }

Hubble time [ASTRON] The reciprocal of the Hubble constant. { 'həb·əl ¸tīm }

hunt [AERO ENG] **1.** Of an aircraft or rocket, to weave about its flight path, as if seeking a new direction or another angle of attack; specifically, to yaw back and forth. **2.** Of a control surface, to rotate up and down or back and forth without being detected by the pilot. { hənt }

hunter's moon [ASTRON] The full moon next following the harvest moon. { 'hən·tərz ¦mün }

Hunting Dogs *See* Canes Venatici. { 'hənt·iŋ ¸dogz }

Hyades [ASTRON] A V-shaped open star cluster about 150 light-years from the sun, which appears in the constellation Taurus near the star Aldebaran. { 'hī·ə¸dēz }

hybrid inflation [ASTRON] A version of the inflationary universe cosmology that relies on the evolution of several interacting fields. { 'hī·brid in¸flā·shən }

hybrid propulsion [AERO ENG] Propulsion utilizing energy released by a liquid propellant with a solid propellant in the same rocket engine. { 'hī·brəd prə'pəl·shən }

hybrid rocket [AERO ENG] A rocket with an engine utilizing a liquid propellant with a solid propellant in the same rocket engine. { 'hī·brəd 'räk·ət }

Hydra [ASTRON] A large constellation of the Southern Hemisphere, right ascension 10 hours, declination 20° south. Also known as Water Monster. { 'hī·drə }

Hydra-Centaurus-Pavo supercluster [ASTRON] The nearest supercluster outside the local supercluster; includes the Centaurus cluster, the Hydra I cluster, and a number of smaller clusters in the constellation Pavo. { ¦hī·drə sen¦tor·əs ¦pä·vō 'sü·pər¸kləs·tər }

Hydra I cluster [ASTRON] A large cluster of galaxies with recession velocities around 3500 kilometers (2200 miles) per second, part of the Hydra-Centaurus-Pavo supercluster. { ¦hī·drə ¸wən 'kləs·tər }

hydrogen burning [ASTROPHYS] Thermonuclear reactions occurring in the cores of main-sequence stars, in which nuclei of hydrogen fuse to form helium nuclei. { 'hī·drə·jən ¸bərn·iŋ }

hydrogen star [ASTROPHYS] A star of spectral class A, a white star with a surface temperature of 8000 to 11,000 K. { 'hī·drə·jən ¸stär }

Hydrus [ASTRON] A southern constellation, right ascension 2 hours, declination 75°S. Also known as Water Snake. { 'hī·drəs }

Hygiea [ASTRON] The fourth largest asteroid, with a diameter of about 260 miles (419 kilometers), mean distance from the sun of 3.14 astronomical units, and C-type surface composition. { hī'jē·ə }

hyperbolic orbit [ASTRON] The path of a body moving along a hyperbola, such as a body that is subject to the gravitational attraction of another body from which it has sufficient energy to escape, and that is otherwise undisturbed. { ¦hī·pər¸bäl·ik 'or·bət }

hypergiant star [ASTRON] A member of the brightest known class of stars, with absolute visual magnitude around −10, about 10^6 times as bright as the sun. { ¦hī·pər¸jī·ənt 'stär }

Hyperion   [ASTRON] A satellite of Saturn approximately 300 miles (480 kilometers) in diameter.   { hī'pir·ē·ən }

hypersonic flight   [AERO ENG] Flight at speeds well above the local velocity of sound; by convention, hypersonic regime starts at about five times the speed of sound and extends upward indefinitely.   { ¦hī·pər'sän·ik 'flīt }

hypersonic glider   [AERO ENG] An unpowered vehicle, specifically a reentry vehicle, designed to fly at hypersonic speeds.   { ¦hī·pər'sän·ik 'glīd·ər }

hypothetical parallax See dynamic parallax.   { ¦hī·pə¦thed·ə·kəl 'par·ə,laks }

Iapetus [ASTRON] A satellite of Saturn that orbits at a mean distance of 2.207×10^6 miles (3.560×10^6 kilometers) and has a diameter of about 900 miles (1500 kilometers). { ,yap·əd·əs }

Icarus [ASTRON] An asteroid with a highly eccentric orbit (eccentricity of 0.827) that crosses the earth's orbit and takes the asteroid to only 0.187 astronomical units from the sun, closer than Mercury. { ik·ə·rəs }

ideal rocket [AERO ENG] A rocket motor or rocket engine that would have a velocity equal to the velocity of its jet gases. { ī'dēl 'räk·ət }

igneous theory See volcanic theory. { 'ig·nē·əs ,thē·ə·rē }

immersion [ASTRON] The disappearance of a celestial body either by passing behind another or passing into another's shadow. { ə'mər·zhən }

impact theory [ASTRON] A theory which holds that most features of the moon's surface were formed by the impact of meteorites. Also known as meteoric theory; meteoritic theory. { 'im,pakt ,thē·ə·rē }

impulsive phase [ASTROPHYS] The first phase of a solar flare, in which x-radiation rises to a maximum in a few seconds or minutes, and can then vary rapidly for several minutes in bursts of decreasing amplitude with rise times as short as 10 milliseconds. { im'pəl·siv ,fāz }

inclination of axis [ASTRON] The angle between a planet's axis of rotation and the perpendicular to the plane of its orbit. { ,iŋ·klə'nā·shən əv 'ak·səs }

inclination of planetary orbits [ASTRON] The angle between the plane of the orbit and the plane of the ecliptic, which is the plane of the earth's orbit. { ,iŋ·klə'nā·shən əv ¦plan·ə,ter·ē 'ȯr·bəts }

inclined orbit [AERO ENG] A satellite orbit which is inclined with respect to the earth's equator. { in'klīnd 'ȯr·bət }

index catalog [ASTRON] A supplement to the New General Catalog of nebulae. { 'in,deks ,kad·əl,äg }

Indian See Indus. { 'in·dē·ən }

Indus [ASTRON] A constellation, right ascension 21 hours, declination 55° south. Also known as Indian. { 'in·dəs }

inertial orbit [ASTRON] The path described by an object that is subject only to gravitational forces, such as a celestial body or a spacecraft that is not under any type of propulsive power. { in'ər·shəl 'ȯr·bət }

infall process [ASTROPHYS] A process in which gas falls upon a very compact object such as a neutron star or black hole, reaching a high velocity and forming a hot plasma; postulated as a model for x-ray sources such as Centaurus X-1 and Hercules X-1. { 'in,fȯl ,prä·səs }

infall zone [ASTRON] The region that forms between the tidal radius of a planet in formation and the actual surface of the planet when the planet contracts from the tidal radius, so that any matter that enters this region falls to the planet's surface. { 'in,fȯl ,zōn }

inferior conjunction [ASTRON] A type of configuration in which two celestial bodies have their least apparent separation; the smaller body is nearer the observer than the larger body, about which it orbits; for example, Venus is closest to the earth at its inferior conjunction. { in'fir·ē·ər kən'jəŋk·shən }

inferior planet   [ASTRON] A planet that circles the sun in an orbit that is smaller than the earth's.   { in'fir·ē·ər 'plan·ət }

inflationary universe cosmology   [ASTRON] A theory of the evolution of the early universe which asserts that at some early time the observable universe underwent a period of exponential expansion, during which the scale of the universe increased by at least 28 orders of magnitude.   { in'flā·shə,ner·ē 'yü·nə,vərs käz'mäl·ə·jē }

inflight start   [AERO ENG] An engine ignition sequence that takes place after takeoff and during flight.   { 'in¦flīt 'stärt }

infrared astronomy   [ASTROPHYS] The study of electromagnetic radiation in the spectrum between 0.75 and 1000 micrometers emanating from astronomical sources.   { ¦in·frə¦red ə'strän·ə·mē }

infrared dwarf See brown dwarf.   { ¦in·frə,red 'dwȯrf }

infrared galaxy   [ASTRON] A galaxy or quasar whose nucleus emits enormous amounts of infrared radiation, in some cases more than 1000 times the output of the entire Milky Way Galaxy at all wavelengths.   { ¦in·frə¦red 'gal·ik·sē }

infrared star   [ASTROPHYS] A star that emits a large amount of radiant energy in the infrared portion of the electromagnetic spectrum.   { ¦in·frə¦red ¦stär }

ingress   [ASTRON] The entrance of the moon into the shadow of the earth in an eclipse, of a planet into the disk of the sun, or of a satellite (or its shadow) onto the disk of the parent planet.   { 'in,gres }

inhibitor   [AERO ENG] A substance bonded, taped, or dip-dried onto a solid propellant to restrict the burning surface and to give direction to the burning process.   { in'hib·əd·ər }

initial mass   [AERO ENG] The mass of a rocket missile at the beginning of its flight.   { i'nish·əl 'mas }

initial mass function   [ASTRON] The distribution of the masses of stars at the time of their formation.   { i'nish·əl 'mas ,fəŋk·shən }

injection   [AERO ENG] The process of placing a spacecraft into a specific trajectory, such as an earth orbit or an encounter trajectory to Mars.   Also known as insertion.   { in'jek·shən }

inner Lagrangian point   [ASTRON] A Lagrangian point that lies between two primary bodies on the line passing through their centers of mass, and through which mass transfer may occur between them.   Also known as conical point.   { ¦in·ər lə'gran·jē·ən ,pȯint }

inner planet   [ASTRON] Any of the four planets (Mercury, Venus, Earth, and Mars) in the solar system whose orbits are closest to the sun.   { ¦in·ər 'plan·ət }

insertion See injection.   { in'sər·shən }

insolation   [ASTRON] **1.** Exposure of an object to the sun. **2.** Solar energy received, often expressed as a rate of energy per unit horizontal surface.   { ,in·sō'lā·shən }

instability strip   [ASTRON] A portion of the Hertzsprung-Russell diagram occupied by pulsating stars; stars traverse this region at least once after they leave the main sequence.   { ,in·stə'bil·əd·ē ,strip }

integrated profile See mean profile.   { 'in·tə,grād·əd 'prō,fīl }

Interamnia   [ASTRON] An asteroid with a diameter of about 203 miles (327 kilometers), mean distance from the sun of 3.06 astronomical units, and F-type (c-like) surface composition.   { ,in·tər'am·nē·ə }

intercalary day   [ASTRON] A day inserted or introduced among others in a calendar, as February 29 during leap years.   { in'tər·kə,ler·ē 'dā }

intercluster medium   [ASTRON] A hot x-ray-emitting gas that pervades the space between the members of a galaxy cluster.   { ,in·tər,kləs·tər 'mēd·ē·əm }

intercombination line   [ASTRON] A spectral line emitted in a transition between energy levels that have different multiplicities, that is, different values of the total spin quantum number.   { ¦in·tər,käm·bə'nā·shən ,līn }

intergalactic   [ASTRON] Pertaining to the space between the galaxies.   { ¦in·tər·gə'lak·tik }

intergalactic matter   [ASTRON] The material between the galaxies.   { ,in·tər·gə¦lak·tik 'mad·ər }

interlocking ring structures [ASTRON] Two lunar craters that overlap, but both have their walls intact, indicating that they were formed at the same time. { |in·tər¦läk· iŋ 'riŋ ¸strək·chərz }

intermediate polar [ASTRON] A member of a class of cataclysmic variable stars whose x-ray and optical light curves display large pulses on time scales of minutes. Also known as DQ Herculis star. { ¸in·tər'mēd·ē·ət 'pōl·ər }

International Celestial Reference Frame [ASTRON] A celestial reference frame made up from the positions of approximately 400 extragalactic radio sources observed with very long baseline interferometry. { ¸in·tər¦nash·ən·əl si¦les·chəl 'ref·rəns ¸frām }

International Celestial Reference System [ASTRON] A system that is realized by the International Celestial Reference Frame, made up from the positions of extragalactic radio sources, and that encompasses the standard reference frames, the transformations between them, and all the constants and motions involved, as well as the time scales specified for the reference frames and origins involved. It was adopted by the International Astronomical Union in 1994 as the fundamental reference system. { ¸in·tər¦nash·ən·əl si¦les·chəl 'ref·rəns ¸sis·təm }

international date line [ASTRON] A jagged arbitrary line, roughly equal to the 180° meridian, where a date change occurs: if the line is crossed from east to west a day is skipped, if from west to east the same day is repeated. { |in·tər¦nash·ən·əl 'dāt ¸līn }

interplanetary dust [ASTRON] Dust particles between the planets. { |in·tər'plan·ə¸ter· ē 'dəst }

interplanetary flight [AERO ENG] Flight through the region of space between the planets, under the primary gravitational influence of the sun. { |in·tər'plan·ə¸ter·ē 'flīt }

interplanetary magnetic field [ASTROPHYS] The magnetic field between the planets. { |in·tər'plan·ə¸ter·ē mag¦ned·ik 'fēld }

interplanetary medium [ASTRON] That part of space containing electromagnetic radiation, dust, gas, and plasma between the planets. { |in·tər'plan·ə¸ter·ē 'mēd·ē·əm }

interplanetary probe [AERO ENG] An instrumented spacecraft that flies through the region of space between the planets. { |in·tər'plan·ə¸ter·ē 'prōb }

interplanetary space [ASTRON] The region that extends beyond near-space away from earth to the other planets in the solar system. { |in·tər'plan·ə¸ter·ē 'spās }

interplanetary spacecraft [AERO ENG] A spacecraft designed for interplanetary flight. { |in·tər'plan·ə¸ter·ē 'spās¸kraft }

interplanetary transfer orbit [AERO ENG] An elliptical trajectory tangent to the orbits of both the departure planet and the target planet. { |in·tər'plan·ə¸ter·ē 'tranz·fər ¸ȯr·bət }

interstellar [ASTRON] Between the stars. { |in·tər¦stel·ər }

interstellar extinction [ASTRON] The dimming of light from stars due to its absorption and scattering by dust grains in the interstellar medium. { |in·tər¦stel·ər ik'stiŋk· shən }

interstellar lines [ASTRON] Dark, narrow lines in the spectra of stars, caused by absorption of radiation by a gaseous medium in space. { |in·tər¦stel·ər 'līnz }

interstellar matter [ASTRON] The gaseous and dust material between the stars. { |in· tər¦stel·ər 'mad·ər }

interstellar probe [AERO ENG] An instrumentated spacecraft propelled beyond the solar system to obtain specific information about interstellar environment. { |in·tər¦stel· ər 'prōb }

interstellar space [ASTRON] The space between the stars. { |in·tər¦stel·ər 'spās }

interstellar travel [AERO ENG] Space flight between stars. { |in·tər¦stel·ər 'trav·əl }

intracluster medium [ASTRON] A hot, tenuous gas that fills the space between the members of a cluster of galaxies and emits x-rays. { ¸in·trə'kləs·tər ¸mēd·ē·əm }

intrinsic luminosity [ASTROPHYS] The total amount of radiation emitted by a star over a specified range of wavelengths. { in'trin·sik ¸lü·mə'näs·əd·ē }

intrinsic variable star [ASTRON] A star that is variable not because of an eclipse. { in'trin·sik ¦ver·ē·ə·bəl 'stär }

invariant plane [ASTRON] The plane that is perpendicular to the total angular momentum of the solar system and passes through its center of mass. { in'ver·ē·ənt 'plān }

Io [ASTRON] A satellite of Jupiter; its diameter is 2300 miles (3700 kilometers). Also known as Jupiter I. { 'ī,ō }

ion engine [AERO ENG] An engine which provides thrust by expelling accelerated or high-velocity ions; ion engines using energy provided by nuclear reactors are proposed for space vehicles. { 'ī,än ,en·jən }

ionization front [ASTRON] A transition region that separates interstellar gas in which a given atomic species (usually hydrogen) is mostly ionized from interstellar gas in which it is mostly neutral. { ,ī·ə·nə'zā·shən ,frənt }

ion propulsion [AERO ENG] Vehicular motion caused by reaction from the high-speed discharge of a beam of electrically charged minute particles, usually positive ions, that are accelerated in an electrostatic field and ejected behind the vehicle. { 'ī,än prə'pəl·shən }

Io torus [ASTRON] A doughnut-shaped region of dense plasma that orbits Jupiter at the radial distance of the satellite Io and results from ionization by solar ultraviolet radiation of gases emitted from Io in volcanic eruptions. { 'ī,ō 'tȯr·əs }

iron meteorite [ASTRON] A type of meteorite that consists mainly of iron and nickel and is several times heavier than any ordinary rock. { 'ī·ərn 'mēd·ē·ə,rīt }

irregular cluster [ASTRON] A type of galaxy cluster that has an overall amorphous appearance, usually showing little overall symmetry or central concentration and often composed of several distinct clumps of galaxies. { i'reg·yə·lər 'kləs·tər }

irregular galaxy [ASTRON] A galaxy which shows no definite order or shape, except that of a general flattened appearance. { i'reg·yə·lər 'gal·ik·sē }

irregular variable star [ASTRON] A star with no fixed period. { i'reg·yə·lər ¦ver·ē·ə·bəl 'stär }

isotropic universe [ASTRON] A universe postulated to have the same properties when viewed from all directions. { ¦ī·sə¦trä·pik 'yü·nə,vərs }

J

Jacobi ellipsoid [ASTROPHYS] A triaxial ellipsoid that can be formed by the surface of a homogeneous, self-gravitating body rotating uniformly with sufficient high angular velocity. { jä'kō·bē i'lip,sóid }

jansky [ASTROPHYS] A unit of measurement of flux density, in units of watt · meter^{-2} · hertz^{-1}; 1 jansky is 10^{-26} W · m^{-2} · Hz^{-1}. Abbreviated Jy. { 'jans·kē }

Janus [ASTRON] A satellite of Saturn which orbits at a mean distance of 151,000 kilometers (94,000 miles) and has an irregular shape with an average diameter of 190 kilometers (120 miles). { 'jā·nəs }

Jeans flux [ASTRON] For a particular constituent of a planetary atmosphere, the number of atoms or molecules that escape from the atmosphere, per unit area per unit time, by virtue of their thermal motions. { 'jēnz ,fləks }

Jeans length [ASTROPHYS] A critical length such that oscillations in homogeneous, infinite media with wavelengths greater than this length are gravitationally unstable. { 'jēnz ,leŋkth }

jet [ASTRON] A narrow, elongated feature in the radio or optical map of an active galaxy, quasar, or object in the Milky Way Galaxy, believed to represent an energetic outflow of gas from a compact astronomical object. { jet }

jet stream [AERO ENG] The stream of gas or fluid expelled by any reaction device, in particular the stream of combustion products expelled from a jet engine, rocket engine, or rocket motor. { 'jet ,strēm }

Johnson-Morgan system *See* UBV system. { 'jän·sən 'mȯr·gən ,sis·təm }

Jovian planet [ASTRON] Any of the four major planets (Jupiter, Saturn, Uranus, and Neptune) that are at a greater distance from the sun than the terrestrial planets (Mercury, Venus, Earth, and Mars). { 'jō·vē·ən ¦plan·ət }

Jovian Van Allen belts [ASTROPHYS] The extended belts of high-energy charged particles that are trapped in Jupiter's magnetic field and cause the microwave nonthermal emission of radio waves observed in the band from about 3 to 70 centimeters. { 'jō·vē·ən ,van 'al·ən ,belts }

Julian calendar [ASTRON] A calendar (replaced by the Gregorian calendar) in which the year was 365.25 days, with the fraction allowing for an extra day every fourth year (leap year); there were 12 months, each 30 or 31 days except for February which had 28 days or in leap year 29. { 'jül·yən 'kal·ən·dər }

Julian date [ASTRON] The sum of the Julian day number and the fraction of a day elapsed since the previous noon. { 'jül·yən ¦dāt }

Julian day [ASTRON] The number of each day, as reckoned consecutively since the beginning of the present Julian period on January 1, 4713 B.C.; it is used primarily by astronomers to avoid confusion due to the use of different calendars at different times and places; the Julian day begins at noon, 12 hours later than the corresponding civil day. { 'jül·yən ¦dā }

Julian ephemeris century [ASTRON] The unit of ephemeris time (ET) in Simon Newcomb's formulas which relate the orbital position of the earth to ephemeris time; the Julian ephemeris century is subdivided into 36,525 days, and 1 ephemeris day = 86,400 ephemeris seconds. { 'jül·yən ə'fem·ə·rəs 'sen·chə·rē }

Juliet [ASTRON] A satellite of Uranus orbiting at a mean distance of 39,990 miles (64,360 kilometers) with a period of 11 hours 52 minutes, and with a diameter of about 52 miles (84 kilometers). { 'jül·ē·ət }

June solstice [ASTRON] Summer solstice in the Northern Hemisphere. { 'jün 'säl·stəs }

Juno [ASTRON] An asteroid with a diameter of about 150 miles (242 kilometers), mean distance from the sun of 2.67 astronomical units, and S-type surface composition; the third asteroid discovered, it is sometimes grouped with Ceres, Pallas, and Vesta as the Big Four, although it is about 14 in size rank. { 'jü·nō }

Jupiter [ASTRON] The largest planet in the solar system, and the fifth in order of distance from the sun; semimajor axis = 485×10^6 miles (780×10^6 kilometers); sidereal revolution period = 11.86 years; mean orbital velocity = 8.2 miles per second (13.2 kilometers per second); inclination of orbital plane to ecliptic = 1.03; equatorial diameter = 88,700 miles (142,700 kilometers); polar diameter = 82,800 miles (133,300 kilometers); mass = about 318.4 (earth = 1). { 'jü·pəd·ər }

Jupiter I–XVII [ASTRON] The 17 known satellite of Jupiter. They are also named: I, Io; II, Europa; III, Ganymede; IV, Callisto; V, Amalthea; VI, Himalia; VII, Elara; VIII, Pasiphae; IX, Sinope; X, Lysithea; XI, Carme; XII, Ananke; XIII, Leda; XIV, Thebe; XV, Adrastea; XVI, Metis; and XVII, 1991 J1. { 'jü·pəd·ər 'wən thrü 'sev·en,tēn }

Jupiter trojan *See* Trojan asteroid. { ¦jü·pəd·ər 'trō·jən }

K

Kapetyn selected areas [ASTRON] Certain areas in the Milky Way Galaxy that the astronomer J.C. Kapetyn suggested be studied intensively in order to determine the structure of the galaxy. Also known as selected areas. { 'kap·əd·ən si¦lek·təd 'er·ē·əz }

Kapetyn's star [ASTRON] A star 13.0 light-years from the solar system, absolute magnitude 11.2, spectrum type M0; has a large proper motion. { 'kap·əd·ənz ¸stär }

K corona [ASTRON] The inner portion of the sun's corona, having a continuous spectrum caused by electron scattering. { 'kā kə¸rō·nə }

Keel *See* Carina. { kēl }

Kelvin-Helmholtz contraction [ASTROPHYS] A contraction of a star once it is formed and before it is hot enough to ignite its hydrogen; the contraction converts gravitational potential energy into heat, some of which is radiated, with the remainder used to raise the internal temperature of the star. { 'kel·vən 'helm¸hōlts kən¸trak·shən }

Kelvin time scale [ASTROPHYS] The time that would be required for a star to contract gravitationally from infinity to its present radius solely through radiation of thermal energy. Also known as thermal time scale. { 'kel·vən 'tīm ¸skāl }

Keplerian ellipse *See* Keplerian orbit. { ke'plir·ē·ən i'lips }

Keplerian motion [ASTRON] Orbital movement of a body about another that is not disturbed by the presence of a third celestial body. { ke'plir·ē·ən 'mō·shən }

Keplerian orbit [ASTRON] An elliptical orbit of a celestial body about another, the latter at a focus of the ellipse. Also known as Keplerian ellipse. { ke'plir·ē·ən 'ȯr·bət }

Kepler's equations [ASTRON] The mathematical relationship between two different systems of angular measurements of the position of a body in an ellipse. { 'kep·lərz i'kwā·zhənz }

Kepler's laws [ASTRON] Three laws, determined by Johannes Kepler, that describe the motions of planets in their orbits: the orbits of the planets are ellipses with the sun at a common focus; the line joining a planet and the sun sweeps over equal areas during equal intervals of time; the squares of the periods of revolution of any two planets are proportional to the cubes of their mean distances from the sun. { 'kep·lərz 'lȯz }

Kepler's supernova [ASTRON] A supernova that appeared in the constellation Ophiuchus in October 1604 and was visible until March 1606. { 'kep·lərz 'sü·pər¸nō·və }

kiloparsec [ASTRON] A distance of 1000 parsecs (3260 light-years). { ¦kil·ə'pär¸sek }

Kirkwood gaps [ASTRON] Regions in the main zone of asteroids where almost no asteroids are found. { 'kərk¸wu̇d ¸gaps }

Kleinmann-Low Nebula [ASTRON] A cool, extended source of infrared radiation in the Orion Nebula, probably a collapsing cloud of gas containing embedded protostars. Abbreviated KL Nebula. { 'klīn¸män 'lō 'neb·yə·lə }

Klein's hypothesis [ASTRON] A theory of the overall structure of the universe that regards the visible universe as part of a large but finite astronomical system called a metagalaxy, which may itself belong to a much larger bounded system. { 'klīnz hī¸päth·ə·səs }

KL Nebula *See* Kleinmann-Low Nebula. { 'kā 'el 'neb·yə·lə }

Kochab [ASTRON] The brighter of the two stars called the Guardian of the Pole in the constellation Ursa Minor. { 'kä·¸käb }

Kohoutek's comet [ASTRON] A comet that was discovered on March 7, 1973, at a distance of 4 astronomical units from the sun, and reached a perihelion of less than 0.1 astronomical unit at the end of 1973. Also known as Comet Kohoutek. { kə'hō·teks ‚käm·ət }

K ratio [AERO ENG] The ratio of propellant surface to nozzle throat area. { 'ka ‚rā·shō }

K star [ASTRON] A star of spectral type K, a cool orange to red star with a surface temperature of about 3600–5000 K (6000–8500°F), and a spectrum resembling that of sunspots in which hydrogen lines have been greatly weakened. { 'kā ‚stär }

Kuiper Belt [ASTRON] A vast reservoir of icy bodies in the region of the solar system beyond the orbit of the planet Neptune, with diameters estimated in the 100–1200-kilometer (60–750-mile) range. Also known as Edgeworth-Kuiper belt. { 'kī·pər ‚belt }

Kuiper Belt dust [ASTRON] A disk-shaped cloud of dust that is believed to be produced by collisions between members of the Kuiper Belt. { 'kī·pər ‚belt ‚dəst }

Lacaille's constellations [ASTRON] The 14 southern constellations identified by N. L. de Lacaille in 1763: Antlia, Callum, Circinus, Crux, Fornax, Horologium, Mensa, Microscopium, Norma, Octans, Pictor, Reticulum, Sculptor, and Telescopium. { lə'kāz ˌkän·stə'lā·shənz }

Lacerta [ASTRON] A small northern constellation lying between Cygnus and Andromeda, and adjoining the northern boundary of Pegasus. Also known as Lizard. { lə'sərd·ə }

Lagoon Nebula [ASTRON] A patchy, luminous gaseous nebula that appears to be surrounded by a much larger region of cold, neutral hydrogen. { lə'gün 'neb·yə·lə }

Lagrangian points [ASTRON] Five points in the orbital plane of two massive objects orbiting about a common center of gravity at which a third object of negligible mass can remain in equilibrium; three points of instable equilibrium are located on the line passing through the centers of mass of the two bodies, and two points of stable equilibrium are located in the orbit of the less massive body, 60° ahead of or behind it. { lə'grän·jē·ən ˌpoins }

lander [AERO ENG] A spacecraft that is designed to land on a celestial body. { 'lan·dər }

Lane-Emden equation See Emden equation. { 'lān 'em·dən iˌkwā·zhən }

Lane-Emden function See Emden function. { 'lān 'em·dən ˌfəŋk·shən }

Lane's law [ASTROPHYS] For the contraction of a star that is assumed to be a sphere of perfect gas, the law that the temperature of the perfect-gas sphere is inversely proportional to its radius. { 'lānz ˌlȯ }

Large Magellanic Cloud [ASTRON] An irregular cloud of stars in the constellation Doradus; it is 160,000 light-years away and nearly 30,000 light-years in diameter. Abbreviated LMC. Also known as Nubecula Major. { 'lärj ¦maj·ə¦lan·ik 'klaúd }

Larissa [ASTRON] A satellite of Neptune orbiting at a mean distance of 45,700 miles (73,600 kilometers) with a period of 13.3 hours, and with a diameter of about 120 miles (190 kilometers). { lə'ris·ə }

last quarter [ASTRON] The phase of the moon at western quadrature, half of the illuminated hemisphere being visible from the earth; has the characteristic half-moon shape. { 'last 'kwȯrd·ər }

late-type star [ASTROPHYS] A star with relatively low surface temperature, in spectral class K or M. { 'lāt ˌtīp ˌstär }

launch [AERO ENG] **1.** To send off a rocket vehicle under its own rocket power, as in the case of guided aircraft rockets, artillery rockets, and space vehicles. **2.** To send off a missile or aircraft by means of a catapult or by means of inertial force, as in the release of a bomb from a flying aircraft. **3.** To give a space probe an added boost for flight into space just before separation from its launch vehicle. { lȯnch }

launch complex [AERO ENG] The composite of facilities and support equipment needed to assemble, check out, and launch a rocket vehicle. { 'lȯnch ˌkäm,pleks }

launching angle [AERO ENG] The angle between the horizontal plane and the longitudinal axis of a rocket or missile at the time of launching. { 'lȯn·chiŋ ˌaŋ·gəl }

launching site [AERO ENG] **1.** A site from which launching is done. **2.** The platform, ramp, rack, or other installation at such a site. { 'lȯn·chiŋ ˌsīt }

launch pad [AERO ENG] The load-bearing base or platform from which a rocket vehicle is launched. Also known as launching pad; pad. { 'lȯnch ˌpad }

launch vehicle [AERO ENG] A rocket or other vehicle used to launch a probe, satellite, or the like. Also known as booster. { 'lȯnch ,ve·ə·kəl }

launch window [AERO ENG] The time period during which a spacecraft or missile must be launched in order to achieve a desired encounter, rendezvous, or impact. { 'lȯnch ,win·dō }

law of equal areas [ASTRON] The second of Kepler's laws. { 'lȯ əv ¦ē·kwəl 'er·ē·əz }

leap year [ASTRON] A year with 366, and not 365, days. { 'lēp ,yir }

Leda [ASTRON] A small satellite of Jupiter with a diameter of about 10 miles (16 kilometers), orbiting at a mean distance of 6.88 × 10⁶ miles (1.11 × 10⁷ kilometers). Also known as Jupiter XIII. { 'lēd·ə }

lenticular galaxy [ASTRON] A galaxy of type S0, consisting of a nucleus surrounded by a disklike structure without arms, and containing little gas and few if any young stars. { len'tik·yə·lər 'gal·ik·sē }

Leo [ASTRON] A northern constellation, right ascension 11 hours, declination 15° north. Also known as Lion. { 'lē·ō }

Leo Minor [ASTRON] A northern constellation, right ascension 10 hours, declination 35° north. Also known as Lesser Lion. { 'lē·ō 'mīn·ər }

Leonids [ASTRON] A meteor shower, the radiant of which lies in the constellation Leo; it is visible between November 10 and 15. { 'lē·ə·nədz }

Leo I system [ASTRON] A dwarf elliptical galaxy about 890,000 light-years distant, having a diameter of about 5000 light-years, and a luminosity about 5 × 10⁶ that of the sun. { 'lē·ō 'wən ,sis·təm }

Leo II system [ASTRON] A dwarf elliptical galaxy about 700,000 light-years distant, having a diameter of about 5000 light-years and a luminosity about 600,000 times that of the sun. { 'lē·ō 'tü ,sis·təm }

lepton era [ASTRON] The period in the early universe, following the hadron era, during which electrons, positrons, neutrinos, and photons were present in nearly equal numbers; roughly between 10⁻⁴ and 20 seconds after the big bang. { 'lep,tän ,ir·ə }

Lepus [ASTRON] A southern constellation, right ascension 5.5 hours, declination 20°S. Also known as Hare. { 'lē·pəs }

Lesser Dog *See* Canis Minor. { 'les·ər ¦dȯg }

Lesser Lion *See* Leo Minor. { 'les·ər ¦lī·ən }

level error [ASTRON] **1.** The difference between the apparent altitude of a celestial object above the apparent horizon and its true altitude above the celestial horizon. **2.** The angle between the east-west mechanical axis of a transit telescope and the horizontal plane. { 'lev·əl ,er·ər }

Lexell's Comet [ASTRON] A small comet that approached to within 2,000,000 miles (3,200,000 kilometers) of earth in 1770; it has not been seen since. { 'lek·selz ,käm·ət }

Libra [ASTRON] A southern constellation, right ascension 15 hours, declination 15° south. Also known as Balance. { 'lē·brə }

libration in latitude *See* lunar libration. { lī'brā·shən in 'lad·ə,tüd }

lifting body [AERO ENG] A maneuverable, rocket-propelled, wingless craft that can travel both in the earth's atmosphere, where its lift results from its shape, and in outer space, and that can land on the ground. { 'lift·iŋ ,bäd·ē }

lifting reentry [AERO ENG] A reentry into the atmosphere by a space vehicle where aerodynamic lift is used, allowing a more gradual descent, greater accuracy in landing at a predetermined spot; it can accommodate greater errors in the guidance system and greater temperature control. { 'lift·iŋ rē'en·trē }

lifting reentry vehicle [AERO ENG] A space vehicle designed to utilize aerodynamic lift upon entering the atmosphere. { 'lift·iŋ rē'en·trē ,vē·ə·kəl }

lift-off [AERO ENG] The action of a rocket vehicle as it leaves its launch pad in a vertical ascent. { 'lif,tȯf }

light curve [ASTROPHYS] A graph showing the variations in brightness of a celestial object; the stellar magnitude is usually shown on the vertical axis, and time is the horizontal coordinate. { 'līt ,kərv }

light cylinder *See* velocity-of-light cylinder. { 'līt ,sil·ən·dər }

light-day [ASTRON] The distance traveled by light in 1 day in a vacuum. { 'līt,dā }

light-echo model [ASTRON] An explanation of superluminal motion wherein an outburst from the center of a quasar illuminates regions at successively greater radii. { 'līt ,ek·ō ,mäd·əl }

light-hour [ASTRON] The distance traveled by light in 1 hour in a vacuum. { 'līt,aú·ər }

lighthouse model [ASTRON] An explanation of superluminal motion wherein a rotating beam from a quasar illuminates stationary clouds that reflect the radio waves to earth, analogous to a lighthouse beacon. { 'līt,haús ,mäd·əl }

light radius See velocity-of-light radius. { 'līt ,rād·ē·əs }

light ratio [ASTROPHYS] A number (2.512) that expresses the ratio of a star's light to that of another star that is one magnitude fainter or brighter. { 'līt ,rā·shō }

light time [ASTRON] The time required for light to travel from a distant object to the earth. { 'līt ,tīm }

light-year [ASTROPHYS] A unit of measurement of astronomical distance; it is the distance light travels in one sidereal year and is equivalent to 9.461 × 10^{12} kilometers or 5.879 × 10^{12} miles. { 'līt ,yir }

lilliputian star See brown dwarf. { ,lil·ə,pyü·shən 'stär }

limb [ASTRON] The circular outer edge of a celestial body; the half with the greater altitude is called the upper limb, and the half with the lesser altitude, the lower limb. { limb }

limb brightening [ASTRON] An observed increase in the intensity of radio, extreme ultraviolet, or x-radiation from the sun or another star from its center to its limb. { 'lim ,brīt·ən·iŋ }

limb darkening [ASTROPHYS] An observed darkening near the surface of the sun's limb as compared to its brighter center. { 'lim ,där·kə·niŋ }

lineament [ASTRON] A prominent linear feature on the lunar surface. { 'lin·ē·ə·mənt }

linear aerospike engine [AERO ENG] A variant of the aerospike engine with a plug nozzle that is V-shaped rather than axisymmetric. { ‚lin·ē·ər 'er·ō,spīk ,en·jən }

line displacement [ASTROPHYS] Widening or shifting of spectral lines of celestial objects arising from several causes, such as gas under high pressure. { 'līn di,splās·mənt }

line of apsides [ASTRON] **1.** The line connecting the two points of an orbit that are nearest and farthest from the center of attraction, as the perigee and apogee of the moon or the perihelion and aphelion of a planet. **2.** The length of this line. { 'līn əv 'ap·sə,dēz }

line profile [ASTROPHYS] A curve that indicates the internal variation in intensity of a spectral line of a celestial body. { 'līn ,prō,fīl }

Lion See Leo. { 'lī·ən }

lithium star [ASTRON] A peculiar giant star of spectral type G or M whose spectrum displays a high abundance of lithium. { 'lith·ē·əm ,stär }

Little Bear See Ursa Minor. { 'lid·əl 'ber }

Little Dipper See Ursa Minor. { 'lid·əl 'dip·ər }

Little Fox See Vulpecula. { 'lid·əl 'fäks }

Little Horse See Equuleus. { 'lid·əl 'hórs }

Little Ruler See Regulus. { 'lid·əl 'rül·ər }

Lizard See Lacerta. { 'liz·ərd }

LMC See Large Magellanic Cloud.

LMXRB See low-mass x-ray binary.

local apparent noon [ASTRON] Twelve o'clock local apparent time, or the instant the apparent sun is over the upper branch of the local meridian. { 'lō·kəl ə¦par·ənt ¦nün }

local apparent time [ASTRON] The arc of the celestial equator, or the angle at the celestial pole, between the lower branch of the local celestial meridian and the hour circle of the apparent or true sun, measured westward from the lower branch of the local celestial meridian through 24 hours. { 'lō·kəl ə¦par·ənt ¦tīm }

local arm See Orion arm. { 'lō·kəl 'ärm }

local civil time [ASTRON] United States terminology during 1925–1952 for local mean time. { 'lō·kəl ¦siv·əl ¦tīm }

local cluster of stars See local star system. { 'lō·kəl ¦kləs·tər əv ¦stärz }

Local Group [ASTRON] A group of at least 20 known galaxies in the vicinity of the sun; the Andromeda Spiral is the largest of the group, and the Milky Way Galaxy is the second largest. { 'lō·kəl 'grüp }

local hour angle [ASTRON] Angular distance west of the local celestial meridian. { 'lō·kəl 'aur ˌaŋ·gəl }

local lunar time [ASTRON] The arc of the celestial equator, or the angle at the celestial pole, between the lower branch of the local celestial meridian and the hour circle of the moon, measured westward from the lower branch of the local celestial meridian through 24 hours; local hour angle of the moon, expressed in time units, plus 12 hours; local lunar time at the Greenwich meridian is called Greenwich lunar time. { 'lō·kəl ¦lü·nər ¦tīm }

local mean noon [ASTRON] Twelve o'clock local mean time, or the instant the mean sun is over the upper branch of the local meridian; local mean noon at the Greenwich meridian is called Greenwich mean noon. { 'lō·kəl ¦mēn ¦nün }

local mean time [ASTRON] The arc of the celestial equator, or the angle at the celestial pole, between the lower branch of the local celestial meridian and the hour circle of the mean sun, measured westward from the lower branch of the local celestial meridian through 24 hours. { 'lō·kəl ¦mēn ¦tīm }

local meridian [ASTRON] The meridian through any particular position which serves as the reference for local time. { 'lō·kəl mə'rid·ē·ən }

local noon [ASTRON] Noon at the local meridian. { 'lō·kəl 'nün }

local sidereal noon [ASTRON] Zero hour local sidereal time, or the instant the vernal equinox is over the upper branch of the local meridian; local sidereal noon at the Greenwich meridian is called Greenwich sidereal noon. { 'lō·kəl sə¦dir·ē·əl ¦nün }

local sidereal time [ASTRON] The arc of the celestial equator, or the angle at the celestial pole which is between the upper branch of the local celestial meridian and the hour circle of the vernal equinox. { 'lō·kəl sə¦dir·ē·əl ¦tīm }

local standard of rest [ASTRON] A frame of reference in which the velocities of neighboring stars average out to zero. { 'lō·kəl 'stan·dərd əv 'rest }

local star cloud See local star system. { 'lō·kəl 'stär ˌklaud }

local star system [ASTRON] The group of stars of which the sun is a member. Also known as local cluster of stars; local star cloud. { 'lō·kəl 'stär ˌsis·təm }

Local Supercluster [ASTRON] A great flattened system of groups and clusters of galaxies, about 1.5 to 2 × 10⁸ light-years (1.4 to 1.9 × 10²⁴ meters) across, which includes the local group of galaxies and the Virgo Cluster. Also known as Virgo Supercluster. { 'lō·kəl 'sü·pər,kləs·tər }

local time [ASTRON] **1.** Time based upon the local meridian as reference, as contrasted with that based upon a zone meridian, or the meridian of Greenwich. **2.** Any time kept locally. { 'lō·kəl 'tīm }

long-period variable [ASTRON] A variable star with a period from about 100 to more than 600 days. { 'lòŋ ˌpir·ē·əd 'ver·ē·ə·bəl }

look angle [AERO ENG] The elevation and azimuth at which a particular satellite is predicted to be found at a specified time. { 'luk ˌaŋ·gəl }

look-back time [ASTRON] The time in the past at which the light now being received from a distant object was emitted. { 'luk ˌbak ˌtīm }

Loop Nebula [ASTRON] A large, bright gaseous nebula in the Large Magellanic Cloud; its diameter is about 260 light-years. Also known as 30 Doradus; Tarantula. { 'lüp 'neb·yə·lə }

loop of retrogression [ASTRON] A loop in the apparent path of a planet, relative to the stars, that is described when the planet undergoes retrograde motion. { 'lüp əv ˌre·trə'gresh·ən }

low-altitude earth orbit [AERO ENG] An artificial satellite orbit whose altitude is less than about 1000 miles (1600 kilometers) above the earth's surface. Abbreviated LEO. { ˌlō ¦al·tə,tüd 'ərth ˌòr·bət }

lower branch [ASTRON] That half of a meridian or celestial meridian from pole to pole which passes through the antipode or nadir of a place. { 'lō·ər ¦branch }

lower culmination See lower transit. { 'lō·ər ˌkəl·mə'nā·shən }

lower limb [ASTRON] That half of the outer edge of a celestial body having the least altitude. { 'lō·ər 'lim }

lower transit [ASTRON] Transit across the lower branch of the celestial meridian. Also known as lower culmination. { 'lō·ər 'trans·ət }

low-mass x-ray binary [ASTRON] A binary system consisting of a low-mass (typically less than 1 solar mass), late-type star and a neutron star or black hole that accretes material through Roche-lobe overflow, resulting in the emission of relatively soft x-rays. Abbreviated LMXRB. { ¦lō ¦mas ¦eks,rā 'bī,ner·ē }

low-surface-brightness galaxy [ASTRON] A galaxy whose spatial density of stars is so low that it is almost invisible. { ¸lō 'sər·fəs ¸brīt·nəs ¸gal·ik·sē }

low-velocity star [ASTRON] One of the Population I stars in the spiral arms of a galaxy which participate in the galactic rotation, thus exhibiting low velocity with respect to the sun and high velocity with respect to the galactic center. { 'lō və¦läs·əd·ē 'stär }

luminosity classes [ASTRON] A classification of stars in an orderly sequence according to their absolute brightness. { ¸lü·mə'näs·əd·ē ¸klas·əz }

luminosity function [ASTRON] The functional relationship between stellar magnitude and the number and distribution of stars of each magnitude interval. Also known as relative luminosity factor. { ¸lü·mə'näs·əd·ē ¸fəŋk·shən }

luminous blue variable [ASTRON] Any of a small group of high-luminosity, unstable, hot supergiant stars that have irregular eruptions or ejections with greatly enhanced mass outflow (10^{-5} to 10^{-4} solar mass per year). { 'lüm·ən·əs ¦blü ¦ver·ē·ə·bəl }

luminous mass [ASTRON] The mass of a celestial object inferred from its luminosity or the luminosities of its components. { 'lü·mə·nəs 'mas }

luminous nebula [ASTRON] A nebula made bright by radiation from stars in the vicinity. { 'lü·mə·nəs 'neb·yə·lə }

Luna [ASTRON] A name for the moon. { 'lü·nə }

lunabase [ASTRON] The basic rocks that make up the dark portions of the lunar surface. Also known as marebase; marial rocks. { 'lü·nə,bās }

Luna program [AERO ENG] A series of Soviet space probes launched for flight missions to the moon. { 'lü·nə ,prō·grəm }

lunar appulse [ASTRON] An eclipse of the moon in which the penumbral shadow of the earth falls on the moon. Also known as penumbral eclipse. { 'lü·nər 'a,pəls }

lunar atmosphere [ASTROPHYS] The volatile elements postulated to have been present on the moon's surface at one time. { 'lü·nər 'at·mə,sfir }

lunar crater [ASTRON] A crater on the moon's surface. { 'lü·nər 'krād·ər }

lunar crust [ASTRON] The outer layer of the moon. { 'lü·nər 'krəst }

lunar day [ASTRON] The time interval between two successive crossings of the meridian by the moon. { 'lü·nər 'dā }

lunar dust [ASTRON] Small particles adhering to the moon's surface. { 'lü·nər 'dəst }

lunar eclipse [ASTRON] Obscuration of the full moon when it passes through the shadow of the earth. { 'lü·nər i'klips }

lunar ephemeris [ASTRON] A computed list of positions the moon will occupy in the sky on certain dates. { 'lü·nər i'fem·ə·rəs }

lunar excursion module [AERO ENG] A manned spacecraft designed to be carried on top of the Apollo service module and having its own power plant for making a manned landing on the moon and a return from the moon to the orbiting Apollo spacecraft. Abbreviated LEM. Also known as lunar module (LM). { 'lü·nər ik¦skər·zhən ¦maj·ül }

lunar flight [AERO ENG] Flight by a spacecraft to the moon. { 'lü·nər 'flīt }

lunar geology *See* selenology. { 'lü·nər jē'äl·ə·jē }

lunar inequality [ASTRON] Variation in the moon's motion in its orbit, due to attraction by other bodies of the solar system. { 'lü·nər ,in·i'kwäl·əd·ē }

lunar interval [ASTRON] The difference in time between the transit of the moon over the Greenwich meridian and a local meridian; the lunar interval equals the difference between the Greenwich and local intervals of a tide or current phase. { 'lü·nər 'in·tər·vəl }

lunarite [ASTRON] The rocks that make up the bright portions of the lunar surface. { 'lü·nə,rīt }

lunar libration [ASTRON] **1.** The effect wherein the face of the moon appears to swing east and west about 8° from its central position each month.   Also known as apparent libration in longitude.   **2.** The state wherein the inclination of the moon's polar axis allows an observer on earth to see about 59% of the moon's surface.   Also known as libration in latitude.   **3.** The small oscillation with which the moon rocks back and forth about its mean rotation rate.   Also known as physical libration of the moon.   { 'lü·nər lī'brā·shən }

lunar magnetic field [ASTROPHYS] The magnetic field of the moon.   { 'lü·nər mag¦ned·ik 'fēld }

lunar mass [ASTROPHYS] The mass of the moon. { 'lü·nər 'mas }

lunar meteoroid [ASTRON] A meteoric particle before it strikes the moon. { 'lü·nər 'mēd·ē·ə,ròid }

lunar month [ASTRON] The period of revolution of the moon about the earth, especially a synodical month. { 'lü·nər 'mənth }

lunar mountain [ASTRON] A mountain on the moon. { 'lü·nər 'maunt·ən }

lunar node [ASTRON] A node of the moon's orbit. { 'lü·nər 'nōd }

lunar nodule [ASTRON] A rock nodule found on the moon. { 'lü·nər 'näj·ül }

lunar noon [ASTRON] The instant at which the sun is over the upper branch of any meridian of the moon. { 'lü·nər 'nün }

lunar nutation [ASTRON] A nodding motion of the earth's axis caused by the inclination of the moon's orbit to the ecliptic; it can displace the celestial pole by 9 seconds of arc from its mean position and has a period of 18.6 years. { 'lü·nər nü'tā·shən }

lunar orbit [AERO ENG] Orbit of a spacecraft around the moon. { 'lü·nər 'ór·bət }

lunar polarization [ASTROPHYS] Polarization of light by the moon's surface. { 'lü·nər ,pō·lə·rə'zā·shən }

lunar pole [ASTRON] A pole of the moon. { 'lü·nər 'pōl }

lunar probe [AERO ENG] Any space probe launched for flight missions to the moon. { 'lü·nər 'prōb }

lunar rock [ASTRON] Rock found on the moon. { 'lü·nər 'räk }

lunar satellite [AERO ENG] A satellite making one or more revolutions about the moon. { 'lü·nər 'sad·əl,īt }

lunar spacecraft [AERO ENG] A spacecraft designed for flight to the moon. { 'lü·nər 'spās,kraft }

lunar time [ASTRON] **1.** Time based upon the rotation of the earth relative to the moon; it may be designated as local or Greenwich, as the local or Greenwich meridian is used as the reference. **2.** Time on the moon. { 'lü·nər 'tīm }

lunar topology [ASTRON] Topology of the moon. { 'lü·nər tə'päl·ə·jē }

lunar year [ASTRON] A time interval comprising 12 lunar (synodic) months. { 'lü·nər 'yir }

lunation [ASTRON] The time period between two successive new moons. { lü'nā·shən }

lunisolar precession [ASTROPHYS] Precession of the earth's equinox caused by the gravitational attraction of the sun and moon. { ¦lü·nə'sō·lər prē'sesh·ən }

Lupus [ASTRON] A southern constellation lying between Centaurus and Scorpius. Also known as Wolf. { 'lü·pəs }

Lupus Loop [ASTRON] A very old supernova remnant, about 400–600 parsecs distant, that forms an extended source of radio waves and soft x-rays. { 'lü·pəs ,lüp }

Lyot division [ASTRON] The gap between rings B and C of Saturn. Also known as French division. { 'lyō di,vizh·ən }

Lyra [ASTRON] A northern constellation; right ascension 19 hours, declination 40° north; its first-magnitude star, Vega, is a navigational star and the most brilliant star in this part of the sky. { 'lī·rə }

α Lyrae See Vega. { ¦al·fə 'lī·rə }

Lyrids [ASTRON] An important meteor shower occurring about April 22; it is regular and predictable, but not heavy, the hourly rate usually being about 7–10. { 'lī·rədz }

Lysithea [ASTRON] A small satellite of Jupiter with a diameter of about 15 miles (24 kilometers), orbiting at a mean distance of about 7.30×10^6 miles (11.75×10^6 kilometers). Also known as Jupiter X. { lī'sith·ē·ə }

M

macho [ASTRON] A massive object in the halo of the Milky Way Galaxy, detected by the effects of gravitational lensing; such objects are hypothesized to account for the dark matter in the universe. Derived from massive compact halo object. { 'mä·chō }

Maclaurin spheroid [ASTRON] A spheroid formed by the surface of a homogeneous, self-gravitating mass in uniform rotation. { mə'klȯr·ən 'sfir,ȯid }

Maffei system [ASTRON] A cluster of external galaxies that lie very close to the galactic plane in the constellation Perseus and are heavily obscured by galactic dust. { mä'fā·ē ,sis·təm }

Magellanic Clouds [ASTRON] Two irregular clouds of stars that are the nearest galaxies to the galactic system; both the Large and Small Magellanic Clouds are identified as Irregular in the classification of E.P. Hubble. Also known as Nubeculae. { ¦maj·ə¦lan·ik 'klaůdz }

Magellanic Stream [ASTRON] A long, thin inhomogeneous filament of gas which extends 120° from the region between the Magellanic Clouds to a point near the south galactic pole. { ¦maj·ə¦lan·ik 'strēm }

Magellanic System [ASTRON] An envelope of neutral hydrogen that includes both the Magellanic Clouds. { ¦maj·ə¦lan·ik 'sis·təm }

magnetic merging See field line reconnection. { mag'ned·ik 'mər·jiŋ }

magnetic star [ASTRON] A star with an unusually strong magnetic field. { mag'ned·ik 'stär }

magnetogram [ASTRON] An image of the sun showing magnetic fields, obtained using a spectroheliograph that has been modified to use the Zeeman effect. { mag'ned·ə,gram }

magnetohydrodynamic arcjet [AERO ENG] An electromagnetic propulsion system utilizing a plasma that is heated in an electric arc and then adiabatically expanded through a nozzle and further accelerated by a crossed electric and magnetic field. { mag¦nēd·ō,hī·drə·dī'nām·ik 'ärk,jet }

magnitude [ASTRON] The relative luminance of a celestial body; the smaller (algebraically) the number indicating magnitude, the more luminous the body. Also known as stellar magnitude. { 'mag·nə,tüd }

magnitude ratio [ASTRON] The ratio (2.512) of relative brightness of two celestial bodies differing in magnitude by 1.0. { 'mag·nə,tüd ¦rā·shō }

magnitude system [ASTRON] A system for designating the relative brightness of stars when photography is used; emulsions of different color sensitivities, used with color filters, permit measurements of starlight of different wavelengths with corresponding determination of magnitude at these wavelengths. { 'mag·nə,tüd ,sis·təm }

main sequence [ASTRON] The band in the spectrum luminosity diagram which has the great majority of stars; their energy derives from core burning of hydrogen into helium. { 'mān 'sē·kwəns }

main sequence star [ASTRON] **1.** Any of those stars in the smooth curve termed the main sequence in a Hertzsprung-Russell diagram. **2.** See dwarf star. { 'mān ¦sē·kwəns 'stär }

main stage [AERO ENG] **1.** In a multistage rocket, the stage that develops the greatest amount of the thrust, with or without booster engines. **2.** In a single-stage rocket vehicle powered by one or more engines, the period when full thrust (at or above

major planet

90%) is attained. **3.** A sustainer engine, considered as a stage after booster engines have fallen away. { 'mān 'stāj }

major planet [ASTRON] Any of the four planets that are larger than earth: Jupiter, Saturn, Neptune, and Uranus. { 'mā·jər 'plan·ət }

manganese star [ASTRON] A star that has an anomalously high ratio of manganese to iron. { 'maŋ·gə,nēs ,stär }

Manger *See* Praesepe. { 'mān·jər }

manned orbiting laboratory [AERO ENG] An earth-orbiting satellite which contains instrumentation and personnel for continuous measurement and surveillance of the earth, its atmosphere, and space. Abbreviated MOL. { 'mand 'òr·bəd·iŋ 'lab·rə,tòr·ē }

manned spacecraft [AERO ENG] A vehicle capable of sustaining a person above the terrestrial atmosphere. { 'mand 'spās,kraft }

March equinox *See* vernal equinox. { 'märch 'ē·kwə,näks }

mare [ASTRON] **1.** One of the large, dark, flat areas on the lunar surface. **2.** One of the less well-defined areas on Mars. { 'mär·ā, mer }

marebase *See* lunabase. { 'mär·ā,bās }

marial rocks *See* lunabase. { 'mär·ē·əl ,räks }

Mars [ASTRON] The planet fourth in distance from the sun; it is visible to the naked eye as a bright red star, except for short periods when it is near its conjunction with the sun; its diameter is about 4150 miles (6700 kilometers). { märz }

Mars probe [AERO ENG] A United States uncrewed spacecraft intended to be sent to the vicinity of the planet Mars, such as in the Mariner or Viking programs. { 'märz ,prōb }

massive compact halo object *See* macho. { ¦mas·iv ¦käm'pakt 'hā·lō ,äb,jekt }

mass-luminosity relation [ASTROPHYS] A relation between stellar magnitudes and mass of the stars; when the absolute magnitudes of stars are plotted versus the logarithms of their masses, the points fall closely along a smooth curve. { 'mas ,lü·mə'näs·əd·ē ri,lā·shən }

mass ratio [AERO ENG] The ratio of the mass of the propellant charge of the rocket to the total mass of the rocket when charged with the propellant. { 'mas 'rā·shō }

matter era [ASTRON] The period in the evolution of the universe, beginning roughly 10^5 years after the big bang, when the universe had cooled to the point at which electrons and protons were able to form neutral hydrogen atoms, and continuing to the present time, during which matter, in the form of atoms, is dominant over radiation. { 'mad·ər ,ir·ə }

maunder minimum [ASTRON] A period of time from about 1650 to 1710 when the sun did not appear to have sunspots. { 'mòn·dər 'min·ə·məm }

Mayall's object [ASTRON] A peculiar object that consists of a ring, a cigar-shaped galaxy, and a bridge that appears to connect them. { 'mā·òlz ,äb·jikt }

medium-altitude earth orbit [AERO ENG] An artificial-satellite orbit whose altitude is between about 1000 and 8000 miles (1600 and 14,500 kilometers) above the earth's surface. Abbreviated MEO. { ,mēd·ē·əm ¦al·tə,tüd 'ərth ,òr·bət }

mean motion [ASTRON] The speed which a planet or its satellite would have if it were moving in a circular orbit with radius equal to its distance from the sun or a central planet with a period equal to its actual period. { 'mēn 'mō·shən }

mean noon [ASTRON] Twelve o'clock mean time, or the instant the mean sun is over the upper branch of the meridian; it may be either local or Greenwich, depending upon the reference meridian. { 'mēn 'nün }

mean place [ASTRON] The position of a star on the celestial sphere as it would be observed from the center of the sun, referred to the mean celestial equator and celestial equinox for the beginning of the year of observation. { 'mēn 'plās }

mean profile [ASTRON] The waveform of a pulsar's periodic emission, averaged synchronously over several hundred pulses or more. Also known as integrated profile; pulse window. { 'mēn 'prō,fīl }

mean sidereal time [ASTRON] Sidereal time adjusted for nutation, to eliminate slight irregularities in the rate. { 'mēn si'dir·ē·əl 'tīm }

mean solar day [ASTRON] The duration of one rotation of the earth on its axis, with respect to the mean sun; the length of the mean solar day is 24 hours of mean solar time or $24^h\ 03^m\ 56.555^s$ of mean sidereal time. { 'mēn 'sō·lər 'dā }

mean solar second [ASTRON] A unit equal to 1/86,400 of a mean solar day. { 'mēn 'sō·lər 'sek·ənd }

mean solar time [ASTRON] Time that has the mean solar second as its unit, and is based on the mean sun's motion. { 'mēn 'sō·lər 'tīm }

mean sun [ASTRON] A fictitious sun conceived to move eastward along the celestial equator at a rate that provides a uniform measure of time equal to the average apparent time; used as a reference for reckoning time, such as mean time or zone time. { 'mēn 'sən }

mean time [ASTRON] Time based on the rotation of the earth relative to the mean sun. { 'mēn 'tīm }

megaparsec [ASTRON] A unit equal to 1,000,000 parsecs. { ˌmeg·ə'pär‚sek }

M82 galaxy [ASTRON] An active, variable spiral galaxy that exhibits strong emission from its center in the radio band, 10^4 to 10^6 times greater than that of normal spirals, and ejection of gases at speeds up to 620 miles (1000 kilometers) per second. { ¦em ¦ād·ē¦tü 'gal·ik·sē }

Mercury [ASTRON] The planet nearest to the sun; it is visible to the naked eye shortly after sunset or before sunrise when it is nearest to its greatest angular distance from the sun. { 'mər·kyə·rē }

mercury-manganese star [ASTRON] A star of spectral type B8 or B9 that has a variable spectrum displaying excesses of phosphorus, manganese, gallium, strontium, yttrium, zirconium, platinum, and mercury, and lacks a strong global magnetic field. { 'mər·kyə·rē 'maŋ·gə‚nēs 'stär }

meridian [ASTRON] **1.** A great circle passing through the poles of the axis of rotation of a planet or satellite. **2.** See celestial meridian. { mə'rid·ē·ən }

meridian altitude [ASTRON] The altitude of a celestial body when it is on the celestial meridian of the observer, bearing 000° or 180° true. { mə'rid·ē·ən ‚al·tə‚tüd }

meridian angle [ASTRON] Angular distance east or west of the local celestial meridian; the arc of the celestial equator, or the angle at the celestial pole, between the upper branch of the local celestial meridian and the hour circle of a celestial body, measured eastward or westward from the local celestial meridian through 180°, and labeled E or W to indicate the direction of measurement. { mə'rid·ē·ən ‚aŋ·gəl }

meridian angle difference [ASTRON] The difference between two meridian angles, particularly between the meridian angle of a celestial body and the value used as an argument for entering into a table. Also called hour angle difference. { mə'rid·ē·ən ¦aŋ·gəl 'dif·rəns }

meridian observation [ASTRON] Measurement of the altitude of a celestial body on the celestial meridian of the observer, or the altitude so measured. { mə'rid·ē·ən ‚äb·sər'vā·shən }

meridian passage [ASTRON] The passage of a celestial body across an observer's meridian. { mə'rid·ē·ən ‚pas·ij }

meridian photometer [ASTRON] An instrument in which mirrors are used to bring the light from two stars which are at or near the celestial meridian simultaneously, but at different altitudes, to a common focus, to compare their brightness. { mə'rid·ē·ən fə'täm·əd·ər }

meridian transit See transit. { mə'rid·ē·ən ‚tran·zət }

mesogranulation [ASTRON] An intermediate scale of convection on the sun, giving rise to a system of convective cells whose size (2500–3700 miles or 4000–6000 kilometers) and lifetime (3–10 hours) lie between those of granulation and supergranulation. { ˌmez·ō‚gran·yə'lā·shən }

mesogranule [ASTRON] One of the convective cells in the mesogranulation observed on the sun. { ˌmez·ə'gran‚yül }

Messier number [ASTRON] A number by which star clusters and nebulae are listed in Messier's catalog; for example, the Andromeda Galaxy is M 31. { me'syā ˌnəm·bər }

Messier's catalog [ASTRON] A listing of 103 star clusters and nebulae compiled in 1784. { me'syāz 'kad·əl,äg }

metagalaxy [ASTRON] The total assemblage of recognized galaxies; essentially this represents the entire material universe. { ¦med·ə'gal·ik·sē }

metal [ASTRON] In stellar spectroscopy, any element heavier than helium. { 'med·əl }

metal-enhanced star formation [ASTRON] The hypothesis that stars form preferentially from regions with higher-than-average atomic number in a chemically inhomogeneous interstellar medium. { 'med·əl in¦hanst 'stär fȯr,mā·shən }

metal-rich star [ASTRON] A star in which the ratio of metals (elements heavier than helium) to hydrogen is greater than that of the Hyades. { 'med·əl ¦rich 'stär }

meteor [ASTRON] The phenomena which accompany a body from space (a meteoroid) in its passage through the atmosphere, including the flash and streak of light and the ionized trail. { 'mēd·ē·ər }

meteor bumper [AERO ENG] A thin metal shield around a space vehicle designed to dissipate the energy of impacting meteoric particles. { 'mēd·ē·ər ,bəm·pər }

meteoric ionization [ASTROPHYS] Ionization resulting from collisional interactions of a meteoroid and its vaporization products with the air. { ,mēd·ē'ȯr·ik ,ī·ə·nə'zā·shən }

meteoric theory *See* impact theory. { ,mēd·ē'ȯr·ik 'thē·ə·rē }

meteoritic theory *See* impact theory. { ,mēd·ē·ə'rid·ik 'thē·ə·rē }

meteoroid [ASTRON] Any solid object moving in interplanetary space that is smaller than a planet or asteroid but larger than a molecule. { 'med·ē·ə,rȯid }

meteor shower [ASTRON] A number of meteors with approximately parallel trajectories. { 'mēd·ē·ər ,shaù·ər }

meteor stream [ASTRON] A group of meteoric bodies with nearly identical orbits. { 'mēd·ē·ər ,strēm }

Metis [ASTRON] The innermost known satellite of Jupiter, having an orbital radius of 79,510 miles (127,960 kilometers) and dimensions of 37 × 21 miles (60 × 34 kilometers. Also known as Jupiter XVI. { 'mēd·əs }

metonic cycle [ASTRON] A time period of 235 lunar months, or 19 years; after this period the phases of the moon occur on the same days of the same months. { me'tän·ik 'sī·kəl }

microlensing [ASTRON] A phenomenon in which a foreground star acts as a gravitational lens when it happens to pass in front of a background star, causing the background starlight to brighten and bend through a ring-shaped region. { 'mī·krō,lenz·iŋ }

micrometeorite [ASTRON] A very small meteorite or meteoritic particle with a diameter generally less than a millimeter. { ¦mī·krō'mē·dē·ə,rīt }

micrometeorite penetration [AERO ENG] Penetration of the thin outer shell (skin) of space vehicles by small particles traveling in space at high velocities. { ¦mī·krō'mē·dē·ə,rīt ,pen·ə'trā·shən }

micrometeoroid [ASTRON] A very small meteoroid with diameter generally less than a millimeter. { ¦mī·krō'mē·dē·ə,rȯid }

microquasar [ASTRON] An object in the Milky Way Galaxy that contains a black hole or neutron star and emits jets of material that exhibit superluminal motion and resemble those emitted from quasars. { ¦mī·krō 'kwā,zär }

microwave background *See* cosmic microwave radiation. { 'mī·krə,wāv 'bak,graùnd }

midcourse correction [AERO ENG] A change in the course of a spacecraft some time between the end of the launching phase and some arbitrary point when terminal guidance begins. { 'mid,kȯrs kə'rek·shən }

midnight sun [ASTRON] The sun when it is visible at midnight; occurs during the summer in high latitudes, poleward of the circle at which the latitude is approximately equal to the polar distance of the sun. { 'mid,nīt 'sən }

Mikheyeve-Smirnov-Wolfenstein effect *See* MSW effect. { mi·¦kā·yev ¦smēr,nȯf 'vülf·ən,shtīn i,fekt }

Milky Way [ASTRON] The faint band of light which encircles the sky and results from the combined light of the many stars near the plane of our galaxy. { 'mil·kē 'wā }

Milky Way Galaxy [ASTRON] The large aggregation of stars and interstellar gas and dust of which the sun is a member. Also known as Galaxy. { 'mil·kē 'wā 'gal·ik·sē }

millisecond pulsar See fast pulsar. { 'mil·ə,sek·ənd 'pəl·sär }

Mimas [ASTRON] A satellite of Saturn orbiting at a mean distance of 115,300 miles (186,000 kilometers). { 'mē·mäs }

Mimosa See β Crucis. { mə'mō·sə }

minisolar nebula [ASTRON] A rotating flattened cloud of gas and dust from which a giant planet and its satellites formed in a manner similar to the formation of the solar system from the solar nebula. { ,min·ē,sō·lər 'neb·yə·lə }

minitrack [AERO ENG] A satellite tracking system consisting of a field of separate antennas and associated receiving equipment interconnected so as to form interferometers which track a transmitting beacon in the payload itself. { 'min·ē,trak }

minor planet [ASTRON] **1.** Those planets smaller than the earth, specifically Mercury, Venus, Mars, and Pluto. **2.** See asteroid. { 'mīn·ər 'plan·ət }

Mira [ASTRON] The first star recognized to be a periodic variable; has a period of 332 ± 9 days and its spectrum changes from M5e at maximum to M9e at minimum it is the prototype of long-period variable stars. { 'mir·ə }

Miranda [ASTRON] A satellite of Uranus orbiting at a mean distance of 76,880 miles (124,000 kilometers). { mə'ran·də }

Mira-type variable [ASTRON] A member of a class of long-period variable stars whose prototype is the star Mira and that exhibit a periodic change in brightness with a time interval between 100 and 1000 days and a range variation of 2.5 magnitudes or more, but that have some cycles that may be much brighter or fainter than others. { 'mir·ə,tīp 'ver·ē·ə·bəl }

missing mass See dark matter. { ,mis·iŋ 'mas }

mixture ratio [AERO ENG] The ratio of the weight of oxidizer used per unit of time to the weight of fuel used per unit of time. { 'miks·chər ,rā·shō }

MJD See modified Julian date.

MKK system See Morgan-Keenan-Kellman system. { ¦em¦kā'kā ,sis·təm }

MK system [ASTRON] A system of classifying stars in which suffixes are added to the designations of the Harvard-Draper sequence to indicate luminosity, ranging from I for supergiants to VI for subdwarfs and white dwarfs. Also known as MKK system; spectral/luminosity classification; Yerkes system. { ¦em,kä ,sis·təm }

mock moon See paraselene. { 'mäk 'mün }

mock sun See anthelion; paranthelion; parhelion. { 'mäk 'sən }

mock sun ring See parhelic circle. { 'mäk 'sən ,riŋ }

modified Julian date [ASTRON] The Julian date minus 2,400,000.5. Abbreviated MJD. { 'mäd·ə,fīd 'jül·yən 'dāt }

module [AERO ENG] A self-contained unit which serves as a building block for the overall structure in space technology; usually designated by its primary function, such as command module or lunar landing module. { 'mäj·ül }

modulus of distance [ASTRON] The quantity $m - M$, where M is the absolute magnitude of a given star and m is its apparent magnitude. Also known as distance modulus. { 'mäj·ə·ləs əv 'dis·təns }

molecular cloud [ASTRON] A dense cloud of interstellar gas in which molecules have formed in appreciable abundance. { mə'lek·yə·lər 'klaüd }

Molniya orbit [AERO ENG] An earth satellite orbit designed for communications satellite service coverage at high latitudes, with an orbital period of slightly less than 12 hours (semisynchronous orbit), inclination of 63.4°, and high eccentricity (0.722), so that the apogee (where the satellite lingers over the service coverage area) is at 25,000 miles (40,000 kilometers) and the perigee is at only 300 miles (500 kilometers). { 'mōl·nē·ə ,ȯr·bət }

Monoceros [ASTRON] A constellation, right ascension 7 hours, declination 5° south; it has mostly faint stars. { mə'näs·ə·rəs }

Monoceros Loop [ASTRON] A filamentary loop nebula about 1000 parsecs (2×10^{16} miles or 3×10^{16} kilometers) distant, which is the remnant of a supernova that took place 50,000–100,000 years ago. { mə'näs·ə·rəs 'lüp }

Monoceros R2 molecular cloud [ASTRON] A massive rotating gas cloud with an abundance of various molecules that contains a compact H II region and is a region of active star formation. { mə'näs·ə·rəs ¦är¦tü mə'lek·yə·lər klaúd }

monocoque [AERO ENG] A type of construction, as of a rocket body, in which all or most of the stresses are carried by the skin. { 'män·ə,käk }

month [ASTRON] **1.** The period of the revolution of the moon around the earth (sidereal month). **2.** The period of the phases of the moon (synodic month). **3.** The month of the calendar (calendar month). { mənth }

moon [ASTRON] **1.** The natural satellite of the earth. **2.** A natural satellite of any planet. { mün }

moonrise [ASTRON] The crossing of the visible horizon by the upper limb of the ascending moon. { 'mün,rīz }

moonset [ASTRON] The crossing of the visible horizon by the upper limb of the descending moon. { 'mün,set }

moon shot [AERO ENG] The launching of a rocket intended to travel to the vicinity of the moon. { 'mün ,shät }

Morgan-Keenan-Kellman system [ASTRON] An expansion of the Harvard sequence to include luminosity, whereby a roman numeral is appended to the Harvard class to indicate position on the Hertzsprung-Russell diagram: I for supergiant, II for bright giant, III for giant, IV for subgiant, and V for dwarf or main sequence. Abbreviated MKK system. { ¦mòr·gən ¦kēn·ən 'kel·mən ,sis·təm }

morning star [ASTRON] A misnomer given to a planet visible to the naked eye, when it rises before the sun. { 'mòrn·iŋ ,stär }

morning twilight [ASTRON] The period of time between darkness and sunrise. { 'mòrn·iŋ ¦twī,līt }

morphological astronomy [ASTRON] A branch of astronomy in which the forms of celestial objects, such as galaxies, are observed, and an attempt is made to draw conclusions from these observations. { ¦mòr·fə¦läj·ə·kəl ə'strän·ə·mē }

moving cluster [ASTRON] **1.** A star cluster with common motions. **2.** An open star cluster near the sun such that measurements may be made of the individual proper motions of the stars. { 'müv·iŋ 'kləs·tər }

M region [ASTROPHYS] Any of the areas on the surface of the sun that are theoretically responsible for magnetic disturbances on the earth. { 'em ,rē·jən }

M star [ASTRON] A spectral classification for a star whose spectrum is characterized by the presence of titanium oxide bands; M stars have surface temperatures of 3000 K for giants and 3400 K for dwarfs. { 'em ,stär }

MSW effect [ASTROPHYS] An enhancement of neutrino oscillation by matter that is predicted to occur when electron neutrinos travel through the sun. Derived from Mikheyeve-Smirnov-Wolfenstein effect. { ¦em¦es'dəb·ə,yü i,fekt }

multiple star [ASTRON] A system of three or more stars which appear to the naked eye as a single star. { 'məl·tə·pəl 'stär }

multipropellant [AERO ENG] A rocket propellant consisting of two or more substances that are fed separately to the combustion chamber. { ¦məl·tə·prə'pel·ənt }

multiring structure [ASTRON] A formation on the moon's surface consisting of two or more craters within a larger crater. { 'məl·tə,riŋ ,strək·chər }

multistage rocket [AERO ENG] A vehicle having two or more rocket units, each unit firing after the one in back of it has exhausted its propellant; normally, each unit, or stage, is jettisoned after completing its firing. Also known as multiple-stage rocket; step rocket. { 'məl·tē,stāj 'räk·ət }

Musca [ASTRON] A southern constellation, right ascension 12 hours, declination 70°S. Also known as Fly. { 'məs·kə }

Myklestad method [AERO ENG] A method of determining the mode shapes and frequencies of the lateral bending modes of space vehicles, taking into account secondary effects of shear and rotary inertia, in which one imagines masses to be concentrated at a finite number of points along the beam, with elastic properties remaining constant between consecutive mass points. { 'mik·əl,stad ,meth·əd }

N

nadir [ASTRON] That point on the celestial sphere vertically below the observer, or 180° from the zenith. { 'nā·dər }

Naiad [ASTRON] A satellite of Neptune orbiting at a mean distance of 30,000 miles (48,000 kilometers) with a period of 7.1 hours, and a diameter of about 34 miles (54 kilometers). { 'nī,ad }

naked T Tauri star *See* weak-line T Tauri star. { ¦nāk·əd ¦tē ¦tȯr·ē 'stär }

Nasmyth focus [ASTRON] One of two locations in a telescope with an altitude-azimuth mounting, located on the horizontal axis on either side of the telescope structure, to which light can be reflected to come to a focus; at such a focus there is usually an observing platform to carry instrumentation. { 'nā,smith ,fō·kəs }

nautical twilight [ASTRON] The interval of incomplete darkness between sunrise or sunset and the time at which the center of the sun's disk is 12° below the celestial horizon. { 'nȯd·ə·kəl 'twī,līt }

near-earth object [ASTRON] An asteroid or comet whose orbit takes it within 1.3 astronomical units of the sun. { ,nir ¦ərth ,äb,jekt }

near stars [ASTRON] Those stars in the celestial neighborhood of the sun, sometimes taken as those 22 stars within 13 light-years of the sun. { 'nir 'stärz }

nebula [ASTRON] Interstellar clouds of gas or small particles; an example is the Horsehead Nebula in Orion. { 'neb·yə·lə }

nebular hypothesis [ASTROPHYS] A theory, proposed in 1796 by Laplace, supposing that the planets originated from the solar nebula surrounding the proto-sun; as the sun cooled, it contracted, rotated faster, and thus caused a ringlike bulging at the equator; this bulge eventually broke off and formed the planets; Laplace further theorized that the sun and other stars formed from clouds of nebulous matter; the theory in this form is not accepted. { 'neb·yə·lər hī'päth·ə·səs }

nebular lines [ASTROPHYS] The spectral lines formed in the glow of bright nebulae; they arise from forbidden atomic transition which can take place because of the very low pressure in the nebula itself. { 'neb·yə·lər 'līnz }

nebular redshift [ASTROPHYS] A systematic shift observed in the spectra of all distant galaxies; the wavelength shift toward the red increases with the distance of the galaxies from the earth. { 'neb·yə·lər 'red,shift }

nebular transitions [ASTROPHYS] Those electronic transitions for doubly ionized argon and chlorine that yield the nebular lines seen in the spectra of gaseous nebulae. { 'neb·yə·lər tran'zish·ənz }

nebular variable *See* T Tauri star. { 'neb·yə·lər 'ver·ē·ə·bəl }

Nemesis [ASTRON] A hypothetical, undetected, brown-dwarf companion of the sun, in a highly elongated orbit that would cause cometary material in Oort's Cloud to fall toward the inner region of the solar system approximately once every 2.8×10^7 years. { 'nem·ə·səs }

Neptune [ASTRON] The outermost of the four giant planets, and the next to last planet, from the sun; it is 30 astronomical units from the sun, and the sidereal revolution period is 164.8 years. { 'nep·tün }

Nereid [ASTRON] The outermost known satellite of Neptune, orbiting at a mean distance of 3,425,900 miles (5,513,400 kilometers) with a period of 360 days, 3.1 hours, and with a diameter of about 210 miles (340 kilometers). { 'nir·ē·əd }

Nestor [ASTRON] One of a group of asteroids whose period of revolution is approximately equal to that of Jupiter, or about 12 years (it is one of the Trojan planets). { 'nes·tər }

Net *See* Reticulum. { net }

neutral region [ASTRON] A region on the sun's surface where the longitudinal magnetic field nearly vanishes; generally found between regions of opposite polarity. { 'nü·trəl ,rē·jən }

neutrino astronomy [ASTRON] The observation of neutrinos from the sun and from extrasolar astronomical sources. { nü¦trē·nō ə'strän·ə·me }

neutrino telescope [ASTRON] Any device for detecting and determining the directions of extraterrestrial neutrinos, such as the deep underwater muon and neutrino detector. { nü'trē·nō 'tel·ə,skōp }

neutron drip [ASTROPHYS] The rapid increase in the abundance of free neutrons that occurs when matter becomes sufficiently dense that electrons are absorbed into nuclei. { 'nü,trän ,drip }

neutron matter [ASTROPHYS] Degenerate matter such as occurs in neutron stars in which there are 8 to 12 times as many neutrons as protons. { 'nü,trän ,mad·ər }

neutron star [ASTRON] A star that is supposed to occur in the final stage of stellar evolution; it consists of a superdense mass mainly of neutrons, and has a strong gravitational attraction from which only neutrinos and high-energy photons could escape so that the star is invisible. { 'nü,trän ,stär }

new inflationary cosmology [ASTRON] A modification of the original inflationary universe cosmology in which the breaking of grand unified symmetry does not involve a tunneling process and is, instead, analogous to a process in which a ball rolls down from a hill with an extremely flat top. { 'nü in'flā·shə,ner·ē käz'mäl·ə·jē }

new moon [ASTRON] The moon at conjunction, when little or none of it is visible to an observer on the earth because the illuminated side is turned away. { 'nü 'mün }

Ney-Allen nebula [ASTRON] An extended source of infrared radiation in the Trapezium region of Orion which displays intense emission at a wavelength of 10 millimeters. { 'nī 'al·ən 'neb·yə·lə }

N galaxy [ASTRON] A galaxy that has the optical appearance of a strongly concentrated object with a semistellar nucleus that is surrounded by a faint halo or extension. { 'en ,gal·ik·sē }

night [ASTRON] The period of darkness between sunset and sunrise. { nīt }

nimbus *See* halo. { 'nim·bəs }

nitrogen sequence [ASTRON] Wolf-Rayet stars in which nitrogen emission bands dominate the spectrum. { 'nī·trə·jən ,sē·kwəns }

Nix Olympica *See* Olympus Mons. { 'niks ə'lim·pə·kə }

no-atmospheric control [AERO ENG] Any device or system designed or set up to control a guided rocket missile, rocket craft, or the like outside the atmosphere or in regions where the atmosphere is of such tenuity that it will not affect aerodynamic controls. { 'nō ,at·mə'sfir·ik kən'trōl }

nodal line [ASTRON] The line passing through the ascending and descending nodes of the orbit of a celestial body. { 'nōd·əl ,līn }

node [ASTRON] **1.** One of two points at which the orbit of a planet, planetoid, or comet crosses the plane of the ecliptic. **2.** One of two points at which a satellite crosses the equatorial plane of its primary. { nōd }

node cycle [ASTRON] The period of time needed for the regression of the moon's nodes to conclude a circuit of 360° of longitude; approximately equal to 18.61 Julian years. { 'nōd ,sī·kəl }

nodical month [ASTRON] The average period of revolution of the moon about the earth with respect to the moon's ascending node, a period of 27 days 5 hours 5 minutes 35.8 seconds, or approximately 27$^1/_4$ days. Also known as draconic month; draconitic month. { 'näd·ə·kəl 'mənth }

nodical year *See* eclipse year. { 'nōd·i·kəl ,yir }

nonimpinging injector [AERO ENG] An injector used in rocket engines which employs parallel streams of propellant usually emerging normal to the face of the injector. { 'nän·im,pin·jiŋ in'jek·tər }

nonradial pulsation [ASTRON] A stellar oscillation whose phase and amplitude vary with stellar latitude and longitude. { ¦nän,rād·ē·əl pəl'sā·shən }

noon [ASTRON] The instant at which a time reference is over the upper branch of the reference meridian. { nün }

noon interval [ASTRON] The predicted time interval between a given instant, usually the time of a morning observation, and local apparent noon; it is used to predict the time for observing the sun on the celestial meridian. { 'nün 'in·tər·vəl }

Norma [ASTRON] A southern constellation; right ascension 16 hours, declination 50°S. Also known as Rule. { 'nòr·mə }

normal spiral galaxy [ASTRON] A galaxy that has a lens-shaped central portion with two arms that begin to coil in the same plane and in the same fashion immediately upon emerging from opposite sides of it. { 'nòr·məl 'spī·rəl 'gal·ik·sē }

North American Nebula [ASTRON] A cloud of dust and gas in the constellation Cygnus; the density of this gas and dust is possibly a thousand times greater than the average density of interstellar gas; a much denser cloud of dust between the nebula and earth obscures portions of the emission nebula to create the appearance of the "Gulf of Mexico" and the "Atlantic Ocean." { 'nòrth ə'mer·i·kən 'neb·yə·lə }

northbound node See ascending node. { 'nòrth,baund 'nōd }

Northern Cross See Cygnus. { 'nòr·thərn 'kròs }

Northern Crown See Corona Borealis. { 'nòr·thərn 'kraun }

north point [ASTRON] The point on the celestial sphere, due north of the observer, at which the celestial meridian intersects the celestial horizon. { 'nòrth ,pòint }

north polar distance [ASTRON] The angular distance between a celestial object and the north celestial pole. { 'nòrth ¦pō·lər 'dis·təns }

north polar sequence [ASTRON] A list of stars in the vicinity of the north celestial pole whose photographic magnitudes have been measured as accurately as possible, and which are used as a basis for determining the magnitudes of other stars. { 'nòrth ¦pō·lər 'sē·kwəns }

North Polar Spur [ASTRON] A region of radio and soft x-ray emission, having a continuous spectral distribution, that extends from the galactic plane to the north galactic pole; believed to be the remnant of an old supernova. { 'nòrth 'pō·lər 'spər }

north pole [ASTRON] The north celestial pole that indicates the zenith of the heavens when viewed from the north geographic pole. { 'nòrth 'pōl }

North Star See Polaris. { 'nòrth 'stär }

nose cone [AERO ENG] A protective cone-shaped case for the nose section of a missile or rocket; may include the warhead, fusing system, stabilization system, heat shield, and supporting structure and equipment. { 'nōz ,kōn }

nova [ASTRON] A star that suddenly becomes explosively bright, the term is a misnomer because it does not denote a new star but the brightening of an existing faint star. { 'nō·və }

novalike symbiotic [ASTRON] A symbiotic star consisting of a red giant combined with a white dwarf, in which thermonuclear reactions of hydrogen and helium from the red giant accreted in the surface layers of the white dwarf are believed to produce outbursts of energy similar to those observed in a nova. Also known as symbiotic nova. { 'nō·və,līk ,sim·bē'äd·ik }

nova-like variable [ASTRON] A binary star in which one component is a white dwarf and the other is a main-sequence star that overflows its Roche lobe and feeds an accretion disk around the white dwarf, resulting in irregular variability. { ,nō·və,līk 'ver·ē·ə·bəl }

nozzle blade [AERO ENG] Any one of the blades or vanes in a nozzle diaphragm. Also known as nozzle vane. { 'näz·əl ,blād }

nozzle thrust coefficient [AERO ENG] A measure of the amplification of thrust due to gas expansion in a particular nozzle as compared with the thrust that would be exerted if the chamber pressure acted only over the throat area. Also known as thrust coefficient. { 'näz·əl ¦thrəst ,kō·i,fish·ənt }

N star [ASTRON] An obsolete classification for a star in the carbon sequence; has about the same temperature as an M star in the Draper catalog. { 'en ,stär }

Nubeculae *See* Magellanic Clouds. { nü'bek·yə,lē }

Nubecula Major *See* Large Magellanic Cloud. { nü'bek·yə·lə 'mā·jər }

Nubecula Minor *See* Small Magellanic Cloud. { nü'bek·yə·lə 'mīn·ər }

nuclear-electric propulsion [AERO ENG] A system of propulsion utilizing a nuclear reactor to generate electricity which is then used in an electric propulsion system or as a heat source for the working fluid. { 'nü·klē·ər i¦lek·trik prə,pəl·shən }

nuclear-electric rocket engine [AERO ENG] A rocket engine in which a nuclear reactor is used to generate electricity that is used in an electric propulsion system or as a heat source for the working fluid. { 'nü·klē·ər i¦lek·trik 'räk·ət ,en·jən }

nuclear time scale [ASTRON] The time it takes for a star to evolve a significant distance from the main sequence when a certain fraction of the hydrogen in its core has been converted to helium by thermonuclear reactions. { 'nü·klē·ər 'tīm ,skāl }

nucleogenesis [ASTROPHYS] The origin of chemical elements in the universe. { ,nü·klē·ə'jen·ə·səs }

nucleosynthesis [ASTROPHYS] The formation of the various nuclides present in the universe by various nuclear reactions, occurring chiefly in the early universe following the big bang, in the interiors of stars, and in supernovae. { ¦nü·klē·ō'sin·thə·səs }

nucleus [ASTRON] The small permanent body of a comet, believed to have a diameter between one and a few tens of kilometers, and to be composed of water and volatile hydrocarbons. { 'nü·klē·əs }

nutation [ASTRON] A slight, slow, nodding motion of the earth's axis of rotation which is superimposed on the precession of the equinoxes; it is the combination of a number of perturbations (lunar, solar, and fortnightly nutation). { nü'tā·shən }

Nysa [ASTRON] An asteroid whose surface composition may resemble that of the aubrites; it has a diameter of approximately 42 miles (68 kilometers) and a mean distance from the sun of 2.42 astronomical units. { 'nī·sə }

O

OB association [ASTRON] A grouping of very young, very hot massive stars of spectral types O and B that has not had time to disperse. { ¦ō'bē ə,sō·sē'ā·shən }

Oberon [ASTRON] One of the five satellites of Uranus; diameter about 870 miles (1400 kilometers). { 'ō·bə,rän }

oblateness [ASTRON] The distortion from a spherical shape in which the diameter at the equator exceeds that at the poles. { ä'blāt·nəs }

oblique ascension [ASTRON] The arc of the celestial equator, or the angle at the celestial pole, between the hour circle of the vernal equinox and the hour circle through the intersection of the celestial equator and the eastern horizon at the instant a point on the oblique sphere rises, measured eastward from the hour circle of the vernal equinox through 24 hours. { ə'blēk ə'sen·chən }

oblique rotator [ASTRON] A star model in which the axis of the magnetic field does not coincide with the axis of rotation. { ə'blēk 'rō,tād·ər }

oblique sphere [ASTRON] The celestial sphere as it appears to an observer between the equator and the pole, where celestial bodies appear to rise obliquely to the horizon. { ə'blēk 'sfir }

obliquity of the ecliptic [ASTRON] The acute angle between the plane of the ecliptic and the plane of the celestial equator, about 23°27'. { ə'blik·wəd·ē əv thə i'klip·tik }

occultation [ASTRON] The disappearance of the light of a celestial body by intervention of another body of larger apparent size; especially, a lunar eclipse of a star or planet. { ,ä·kəl'tā·shən }

occulting bar [ASTRON] A bar placed in the focal plane of a telescope eyepiece to cover part of the field of view, usually to cover a bright object in order to permit observation of a nearby faint object. { ə'kəlt·iŋ ,bär }

occulting disk [ASTRON] A small metal disk placed in the focal plane of the eyepiece of a telescope, usually to cover a bright object in order to permit observation of a faint one. { ə'kəlt·iŋ ,disk }

Octans [ASTRON] The constellation that includes the south celestial pole. Also known as Octant. { 'äk,tanz }

Octant See Octans. { 'äk·tənt }

Olbers' paradox [ASTRON] If the universe were static, of infinite age, and the galaxies distributed isotropically, the distance attenuation of their light would be exactly balanced by the increase in number in successive spherical shells centered at the earth; hence the night sky would be of daylight brightness instead of dark. { 'ōl·bərz 'par·ə,däks }

old inflationary cosmology [ASTRON] The original version of the inflationary universe cosmology in which a quantum-mechanical tunneling process is responsible for the phase transition of the universe to a state in which grand unified symmetry is broken. { 'ōld in'flā·shə,ner·ē käz'mäl·ə·jē }

Olympus Mons [ASTRON] The largest volcano on Mars; it is approximately 360 miles (600 kilometers) across at its base and stands approximately 16 miles (26 kilometers) above the surrounding terrain. Also known as Nix Olympica. { ə'lim·pəs 'mänz }

Omega Nebula [ASTRON] A bright H II region in the constellation Sagittarius that is both a bright far-infrared source and a double radio source. Also known as Swan Nebula. { ō'meg·ə 'neb·yə·lə }

Oort dark matter

Oort dark matter [ASTRON] Matter of unknown nature that is postulated to exist in the disk of the Milky Way Galaxy in order to account for the spatial and velocity distributions of stars in the direction perpendicular to the galactic plane. { ¦ȯrt 'därk ˌmad·ər }

Oort's Cloud [ASTRON] A cloud of comets at distances from 75,000 to 150,000 astronomical units from the sun, which has been proposed as a source of comets that pass near the sun. { 'ȯrts ˌklau̇d }

open arc [ASTRON] A crater arc in which the craters do not touch each other. { 'ō·pən 'ärk }

open chain [ASTRON] A crater chain in which the craters do not touch each other. { 'ō·pən 'chān }

open cluster [ASTRON] One of the groupings of stars that are concentrated along the central plane of the Milky Way; most have an asymmetrical appearance and are loosely assembled, and the stars are concentrated in their central region; they may contain from a dozen to many hundreds of stars. Also known as galactic cluster. { 'ō·pən 'kləs·tər }

open inflation [ASTRON] A version of the inflationary universe cosmology in which there is sufficient inflation to solve the horizon problem but not enough to result in a flat universe. { ˌō·pən in'flā·shən }

open universe [ASTRON] A cosmological model in which the volume of the universe is infinite and its expansion will continue forever. { 'ō·pən 'yü·nəˌvərs }

Ophelia [ASTRON] A satellite of Uranus orbiting at a mean distance of 33,400 miles (53,760 kilometers) with a period of 9 hours 3 minutes, and a diameter of about 20 miles (32 kilometers); the outer shepherding satellite for the outermost ring of Uranus. { ō'fēl·yə }

Ophiucus [ASTRON] A large constellation centered near right ascension 17 hours, declination 10° south; it includes a portion of the ecliptic between the constellations Scorpius and Sagittarius, although it is not considered a constellation of the zodiac. Also known as Serpent Bearer. { ¦äf·i¦yü·kəs }

Oppenheimer-Volkoff limit [ASTRON] The upper limit on the mass of a neutron star, above which there is no stable equilibrium configuration and it is predicted that matter will collapse into a black hole. { 'äp·ən̩hī·mər 'fȯl̩kȯf ˌlim·ət }

opposition [ASTRON] The situation of two celestial bodies having either celestial longitudes or sidereal hour angles differing by 180°; the term is usually used only in relation to the position of a superior planet or the moon with reference to the sun. { ˌäp·ə'zish·ən }

optical double star [ASTRON] Two stars not formally a physical system but that appear to be a typical double star; a false binary star whose components happen to lie nearby in the same line of sight. { 'äp·tə·kəl 'dəb·əl 'stär }

optical galaxy [ASTRON] One of the galaxies that appear as nearly starlike, generally having compact nuclei. { 'äp·tə·kəl 'gal·ik·sē }

orbital curve [AERO ENG] One of the tracks on a primary body's surface traced by a satellite that orbits about it several times in a direction other than normal to the primary body's axis of rotation; each track is displaced in a direction opposite and by an amount equal to the degrees of rotation between each satellite orbit. { 'ȯr·bəd·əl 'kərv }

orbital decay [AERO ENG] The lessening of the eccentricity of the elliptical orbit of an artificial satellite. { 'ȯr·bəd·əl di'kā }

orbital direction [AERO ENG] The direction that the path of an orbiting body takes; in the case of an earth satellite, this path may be defined by the angle of inclination of the path to the equator. { 'ȯr·bəd·əl di'rek·shən }

orbital node [ASTRON] One of the two points at which the orbit of a planet or satellite crosses the plane of the ecliptic or equator. { 'ȯr·bəd·əl 'nōd }

orbital period [ASTRON] The interval between successive passages of a satellite through the same specified point in its orbit. { 'ȯr·bəd·əl 'pir·ē·əd }

orbital rendezvous [AERO ENG] **1.** The meeting of two or more orbiting objects with zero relative velocity at a preconceived time and place.   **2.** The point in space at which such an event occurs.   { 'ȯr·bəd·əl 'rän·də,vü }

orbital velocity   [ASTRON] The instantaneous velocity at which an earth satellite or other orbiting body travels around the origin of its central force field.   { 'ȯr·bəd·əl və'läs·əd·ē }

Origem Loop   [ASTRON] A loop of gas on the boundary between Orion and Gemini about 60 parsecs (1.2×10^{15} miles or 1.8×10^{15} kilometers) in radius and 1000 parsecs (2×10^{16} miles or 3×10^{16} kilometers) distant, with at least five nebulae embedded in it.   { 'ȯr·ə,jem ,lüp }

Orion   [ASTRON] A northern constellation near the celestial equator, right ascension 5 hours, declination 5° north.   Also known as Warrior.   { ə'rī·ən }

Orion A   [ASTRON] A giant molecular cloud, 100,000 times more massive than the sun, mostly made of molecular hydrogen but best traced in the 2.6-millimeter emission line of ^{13}CO, the rarer isotopic variant of carbon monoxide; the Orion Nebula is located in front of its northern part.   { ə,rī·ən 'ā }

Orion arm   [ASTRON] The spiral arm of the Milky Way Galaxy that has a spur in which the sun is located.   Also known as local arm.   { ə'rī·ən ,ärm }

Orion B   [ASTRON] A giant molecular cloud located in the northern part of the constellation Orion.   { ə,rī·ən 'bē }

Orionids   [ASTRON] A meteor shower seen in October in the northern hemisphere; its radiant lies in the constellation Orion.   { ə'rī·ə,nidz }

Orion molecular clouds   [ASTRON] Two molecular clouds in the Orion Nebula; one has about 300,000 hydrogen molecules per cubic centimeter and contains the Becklin-Nengebauer object and the Kleinmann-Low Nebula, while the other is centered on a cluster of infrared sources.   { ə'rī·ən mə'lek·yə·lər 'klaȯdz }

Orion Nebula   [ASTRON] A luminous cloud surrounding Ori, the northern star in Orion's dagger; visible to the naked eye as a hazy object.   Also known as Great Nebula of Orion.   { ə'rī·ən 'neb·yə·lə }

Orion OB association   [ASTRON] A loose, gravitationally unbound grouping of hot massive stars with spectral types A, B, and O, which originated in the giant molecular clouds Orion A and Orion B, together with tens of thousands of low-mass young stars which formed from the Orion molecular clouds.   { ə,rī·ən ,ō'bē ə,sō·shē,ā·shən }

Orion spur   [ASTRON] That portion of the Orion arm within which the sun is located.   { ə'rī·ən 'spər }

orrery   [ASTRON] A model of the solar system equipped with mechanical devices to make the planets move at their correct relative velocities around the sun.   { 'ȯr·ər·ē }

oscillating universe   [ASTRON] An extension of the closed universe model in which the universe, after contracting toward a singularity, undergoes another big bang to begin a new cycle, and thenceforth oscillates between successive expansions and contractions, each contraction followed by a new big bang.   { 'äs·ə,lād·iŋ 'yü·nə,vərs }

osculating orbit   [ASTRON] The orbit which would be followed by a body such as an asteroid or comet if, at a given time, all the planets suddenly disappeared, and it then moved under the gravitational force of the sun alone.   { 'äs·kyə,lād·iŋ 'ȯr·bət }

O star   [ASTRON] A star of spectral type O, a massive, very hot blue star with a surface temperature of at least 35,000 K (63,000°F), and a spectrum in which lines of singly ionized helium are prominent.   { 'ō ,stär }

Ostriker-Peebles halo   [ASTRON] A spherical distribution of matter of unknown nature that is postulated to exist to account for the stability of the highly flattened visible disk of the Milky Way Galaxy.   { 'äs·trīk·ər 'pēb·əlz ,hā,lō }

O-type star   [ASTRON] A spectral-type classification in the Draper catalog of stars; a star having spectral type O; a very hot, blue star in which the spectral lines of ionized helium are prominent.   { 'ō ,tīp 'stär }

outer planets   [ASTRON] The planets with orbits larger than that of Mars: Jupiter, Saturn, Uranus, Neptune, and Pluto.   { 'aȯd·ər 'plan·əts }

outer space   [ASTRON] A general term for any region that is beyond the earth's atmosphere.   { 'aȯd·ər 'spās }

outgassing [ASTRON] The ejection of gases trapped within a planet so that they are added to the planet's atmosphere. { 'aút,gas·iŋ }

overall efficiency [AERO ENG] The efficiency of a jet engine, rocket engine, or rocket motor in converting the total heat energy of its fuel first into available energy for the engine, then into effective driving energy. { ¦ō·vər¦ól i'fish·ən·sē }

Owl Nebula [ASTRON] A large planetary nebula in Ursa Major which has two large, circular darker areas in an otherwise opaque spherical shell. { 'aúl 'neb·yə·lə }

P

Pacific Standard Time *See* Pacific time. { pə'sif·ik 'stan·dərd 'tīm }

Pacific time [ASTRON] The time for a given time zone that is based on the 120th meridian and is the eighth zone west of Greenwich. Also known as Pacific Standard Time. { pə'sif·ik 'tīm }

pad *See* launch pad. { pad }

pad deluge [AERO ENG] Water sprayed on certain launch pads during rocket launching in order to reduce the temperatures of critical parts of the pad or the rocket. { 'pad 'del,yüj }

paddle [AERO ENG] A large, flat, paddle-shaped support for solar cells, used on some satellites. { 'pad·əl }

Pallas [ASTRON] The second-largest asteroid, with a diameter of about 324 miles (540 kilometers), mean distance from the sun of 2.769 astronomical units, and C-type surface composition. { 'pal·əs }

Pandora [ASTRON] A satellite of Saturn which orbits at a mean distance of 88,000 miles (142,000 kilometers), just outside the F ring; together with Prometheus, it holds this ring in place. { pan'dör·ə }

parabolic flight [AERO ENG] A space flight occurring in a parabolic orbit. { ¦par·ə¦bäl·ik 'flīt }

parabolic orbit [ASTRON] An orbit whose overall shape is like a parabola; the orbit represents the least eccentricity for escape from an attracting body. { ¦par·ə¦bäl·ik 'ör·bət }

parabolic velocity [ASTRON] The velocity attained by a celestial body in a parabolic orbit. { ¦par·ə¦bäl·ik və'läs·əd·ē }

parachute [AERO ENG] **1.** A contrivance that opens out somewhat like an umbrella and catches the air so as to retard the movement of a body attached to it. **2.** The canopy of this contrivance. { 'par·ə,shüt }

parachute-opening shock [AERO ENG] The shock or jolt exerted on a suspended parachute load when the parachute fully catches the air. { 'par·ə,shüt ,ōp·ə·niŋ ,shäk }

paraglider [AERO ENG] A triangular device on a rocket or spacecraft that consists of two flexible sections and resembles a kite; deployed to assist in guiding or landing a spacecraft or in recovering a launching rocket. { 'par·ə,glīd·ər }

parallactic displacement [ASTRON] The apparent changes in the position of a star due to changes in the position of the earth as it moves around the sun. Also known as parallactic shift. { ¦par·ə¦lak·tik di'splās·mənt }

parallactic ellipse [ASTRON] An annual apparent elliptical course of a celestial body on the celestial sphere about its mean position; caused by the elliptical orbital motion of the earth. { ¦par·ə¦lak·tik i'lips }

parallactic equation [ASTRON] An inequality in the moon's motion caused by the sun's perturbing effect on the moon being greater in that half of the moon's apparent orbit around the earth when at new moon rather than at full moon. Also known as parallactic inequality. { ¦par·ə¦lak·tik i'kwā·zhən }

parallactic inequality *See* parallactic equation. { ¦par·ə¦lak·tik ,in·i'kwäl·əd·ē }

parallactic motion [ASTRON] An apparent motion of stars away from the point in the celestial sphere toward which the sun is moving. { ¦par·ə¦lak·tik 'mō·shən }

parallactic orbit [ASTRON] The apparent orbit of a star as it appears to move once around in the sky each year; the motion is caused by the earth's orbital motion around the sun. { ¦par·ə¦lak·tik 'ȯr·bət }

parallactic shift See parallactic displacement. { ¦par·ə¦lak·tik 'shift }

parallax-second See parsec. { 'par·ə‚laks ¦sek·ənd }

parallel of altitude [ASTRON] A circle on the celestial sphere parallel to the horizon connecting all points of equal altitude. Also known as almucantar; altitude circle. { 'par·ə‚lel əv 'al·tə‚tüd }

parallel of declination [ASTRON] A small circle of the celestial sphere parallel to the celestial equator. Also known as celestial parallel; circle of equal declination. { 'par·ə‚lel əv ‚dek·lə'nā·shən }

parallel of latitude See circle of longitude. { 'par·ə‚lel əv 'lad·ə‚tüd }

parallel sphere [ASTRON] The celestial sphere as it appears to an observer at the pole, where celestial bodies appear to move parallel to the horizon. { 'par·ə‚lel 'sfir }

paranthelion [ASTRON] A refraction phenomenon similar to a parhelion, but occurring generally at a distance of 120° (occasionally 90° and 140°) from the sun, on the parhelic circle. Also known as mock sun. { ‚par·ən'thē·lē‚än }

paraselene [ASTRON] A weakly colored lunar halo identical in form and optical origin to the solar parhelion; paraselenae are observed less frequently than are parhelia, because of the moon's comparatively weak luminosity. Also known as mock moon. { ¦par·ə·sə'lēn }

paraselenic circle [ASTRON] A halo phenomenon consisting of a horizontal circle passing through the moon, corresponding to the parhelic circle through the sun, and produced by reflection of moonlight from ice crystals. { ¦par·ə·sə'len·ik 'sər·kəl }

parhelic circle [ASTRON] A halo consisting of a faint white circle passing through the sun and running parallel to the horizon for as much as 360° of azimuth. Also known as mock sun ring. { pär'hē·lik }

parhelion [ASTRON] Either of two colored luminous spots that appear at points 22° (or somewhat more) on both sides of the sun and at the same elevation as the sun; the solar counterpart of the lunar paraselene. Also known as mock sun; sun dog. { pär'hēl·yən }

Parker bound [ASTROPHYS] An upper bound on the density of magnetic monopoles that is obtained from arguments based on the existence of a galactic magnetic field. { ¦pär·kər ¦baúnd }

Parker model [ASTRON] A model of the solar wind that assumes the solar wind is driven by the thermal pressure of the hot coronal gas. { 'pär·kər ‚mäd·əl }

parking orbit [AERO ENG] A temporary earth orbit during which the space vehicle is checked out and its trajectory carefully measured to determine the amount and time of increase in velocity required to send it into a final orbit or into space in the desired direction. { 'pärk·iŋ ‚ȯr·bət }

parsec [ASTRON] The distance at which a star would have a parallax equal to 1 second of arc; 1 parsec equals 3.258 light-years or 3.08572×10^{13} kilometers. Derived from parallax-second. { 'pär‚sek }

partial eclipse [ASTRON] An eclipse in which only part of the source of light is obscured. { 'pär·shəl i'klips }

Pasiphae [ASTRON] A small satellite of Jupiter with a diameter of about 35 miles (56 kilometers), orbiting with retrograde motion at a mean distance of about 1.46×10^7 miles (2.35×10^7 kilometers). Also known as Jupiter VIII. { pə'sif·ə‚ē }

pass [AERO ENG] **1.** A single circuit of the earth made by a satellite; it starts at the time the satellite crosses the equator from the Southern Hemisphere into the Northern Hemisphere. **2.** The period of time in which a satellite is within telemetry range of a data acquisition station. { pas }

patera [ASTRON] A shallow crater with complex scalloped edges, which occurs on Mars but has not been observed on any other planet. { pə'ter·ə }

Patientia [ASTRON] An asteroid with a diameter of about 153 miles (247 kilometers), mean distance from the sun of 3.06 astronomical units, and B-type (C-like) surface composition. { ‚pä·shē'en·chə }

Patroclus group *See* pure Trojan group. { pə'trō·kləs ˌgrüp }

Pavo [ASTRON] A southern constellation; right ascension 20 hours, declination 65°S. Also known as Peacock. { 'pä·vō }

payload [AERO ENG] That which an aircraft, rocket, or the like carries over and above what is necessary for the operation of the vehicle in its flight. { 'pā,lōd }

payload-mass ratio [AERO ENG] Of a rocket, the ratio of the effective propellant mass to the initial vehicle mass. { 'pā,lōd ¦mas ˌrā·shō }

P Cygni star [ASTRON] An explosive variable star of spectral type B, with broad emission lines and strong absorption of violet light. { ¦pē 'sig·nē ˌstär }

Peacock *See* Pavo. { 'pē,käk }

peculiar star [ASTRON] A star that does not fit into a standard spectral classification. { pə'kyül·yər 'stär }

peculiar velocity [ASTRON] Superposed on the systematic rotation of the galaxy are individual motions of the stars; each star moves in a somewhat elliptical orbit and therefore shows a velocity of its own (peculiar velocity) to the local standard of rest, the standard moving in a circular orbit around the galactic center. { pə'kyül·yər və'läs·əd·ē }

Pegasus [ASTRON] A northern constellation; right ascension 22 hours, declination 20°N. Also known as Winged Horse. { 'peg·ə·səs }

penumbra [ASTRON] The outer, relatively light part of a sunspot. { pə'nəm·brə }

penumbral eclipse *See* lunar appulse. { pə'nəm·brəl i'klips }

penumbral waves [ASTRON] Waves that are often observed to propagate outward across the penumbrae of large sunspots when the penumbrae are viewed in the light of the Hα spectral line of hydrogen. { pə'nəm·brəl 'wāvz }

perfect cosmological principle [ASTRON] The assumption that the universe is homogeneous and isotropic, and does not change with time. { 'pər·fikt ˌkäz·mə'läj·ə·kəl 'prin·sə·pəl }

periapsis [ASTRON] The orbital point nearest the center of attraction of an orbiting body. { ˌper·ē'ap·səs }

periastron [ASTRON] The coordinates and time when the two stars of a binary star system are nearest to each other in their orbits. { ¦per·ē¦as·trən }

pericronus [ASTRON] The nearest point of a satellite in its orbit about Saturn. Also known as perisaturnium. { ¦per·ə'krō·nəs }

pericynthion [ASTRON] The point in the orbit of a satellite around the moon that is nearest to the moon. { ˌper·ə'sin·thē,än }

perigalacticon [ASTRON] The point in the orbit of a star that is closest to the center of the Galaxy. { ˌper·i·gə'lak·tē,kän }

perigee [ASTRON] The point in the orbit of the moon or other satellite when it is nearest the earth. { 'per·ə,jē }

perigee-to-perigee period *See* anomalistic period. { 'per·ə,jē tü 'per·ə,jē ˌpir·ē·əd }

perihelion [ASTRON] That orbital point nearest the sun when the sun is the center of attraction. { ¦per·ə¦hēl·yən }

perijove [ASTRON] The nearest point of a satellite in its orbit about Jupiter. { 'per·ə,jōv }

period [ASTRON] The average time interval for a variable star to complete a cycle of its variations. { 'pir·ē·əd }

periodic perturbation [ASTRON] Small deviations from the computed orbit of a planet or satellite; the deviations extend through cycles that generally do not exceed a century. { ¦pir·ē¦äd·ik pər·tər'bā·shən }

period-luminosity relation [ASTRON] Relation between the periods of Cepheid variable stars and their absolute magnitude; the absolutely brighter the star, the longer the period. { 'pir·ē·əd ˌlü·mə'näs·əd·ē ri,lā·shən }

perisaturnium *See* pericronus. { 'per·ə,sə'tər·nē·əm }

perpetual calendar [ASTRON] A table or mechanical device used to determine the day of the week corresponding to any given date over a period of many years. { pər'pech·ə·wəl 'kal·ən·dər }

95

Perseids

Perseids [ASTRON] A meteor shower whose radiant lies in the constellation Perseus; it reaches a maximum about August 12. { 'pər·sē·ədz }

Perseus [ASTRON] A northern constellation; right ascension 3 hours; declination 45°N. { 'pər·sē·əs }

Perseus A [ASTRON] A strong radio source, having a redshift $z = 0.018$ and centered on the Seyfert galaxy NGC 1275, that undergoes extremely violent outbursts. { 'pər·sē·əs 'ā }

Perseus arm [ASTRON] A spiral arm of the Milky Way galaxy visible in the constellation Perseus, located (as viewed from the earth) in the direction opposite that of the galactic center. { 'pər·sē·əs ,ärm }

Perseus cluster [ASTRON] An irregular, diffuse cluster of galaxies centered on the Seyfert galaxy NGC 1275, with redshift $z = 0.018$ { 'pər·sē·əs ,kləs·tər }

Perseus-Pisces supercluster [ASTRON] A dominant supercluster that occurs in the south galactic hemisphere and includes the Perseus cluster toward one end of a long, filamentary central condensation. { ¦pər·sē·əs ¦pī,sēz 'sü·pər,kləs·tər }

Perseus X-1 [ASTRON] The strongest known x-ray source outside the Milky Way galaxy, centered on the Seyfert galaxy NGC 1275. { 'pər·sē·əs 'eks 'wən }

perturbation [ASTRON] A deviation of an astronomical body from its computed orbit because of the attraction of another body or bodies. { ,pər·tər'bā·shən }

phase [ASTRON] One of the cyclically repeating appearances of the moon or other orbiting body as seen from earth. { fāz }

phase quadrature *See* quadrature. { 'fāz ,kwäd·rə·chər }

Phillips relation [ASTRON] A correlation between the relatively small variation in the peak brightness of type Ia supernovae and their rates of decline, whose use enables the distances of type Ia supernovae to be estimated. { 'fil·ips ri,lā·shən }

Phobos [ASTRON] A satellite of Mars; it is the larger of the two satellites, with a diameter of about 15 miles (24 kilometers). { 'fō,bós }

Phoebe [ASTRON] A satellite of the planet Saturn; its diameter is judged to be about 190 miles (320 kilometers); it has an eccentric orbit and retrograde revolution. { 'fē·bē }

Phoenix [ASTRON] A southern constellation; right ascension 1 hour, declination 50°S. { 'fē·niks }

photoelectric magnitude [ASTRON] The magnitude of a celestial object, as measured by a photoelectric photometer attached to a telescope. { ¦fōd·ō·i'lek·trik 'mag·nə,tüd }

photographic magnitude [ASTRON] The magnitude of a star, as obtained by measuring the apparent size of a star's image on a photographic emulsion sensitive to blue light at wavelengths between 400 and 500 nanometers. { ¦fōd·ə¦graf·ik 'mag·nə,tüd }

photographic meteor [ASTRON] A meteor which has been photographed for the purpose of determining its origin, velocity, and other characteristics. { ¦fōd·ə¦graf·ik 'mēd·ē·ər }

photometric binary *See* eclipsing variable star. { ¦fōd·ō¦me·trik 'bī,ner·ē }

photometric parallax [ASTRON] The annual parallax of a star too far away for its parallax to be measured directly, as calculated from its apparent magnitude and its absolute magnitude inferred from its spectral type. { ¦fōd·ə¦me·trik 'par·ə,laks }

photosphere [ASTRON] The intensely bright portion of the sun visible to the unaided eye; it is a shell a few hundred miles in thickness marking the boundary between the dense interior gases of the sun and the more diffuse cooler gases in the outer portions of the sun. { 'fōd·ə,sfir }

photospheric granulation *See* granulation. { ¦fōd·ə¦sfir·ik ,gran·yə'lā·shən }

photovisual magnitude [ASTRON] The magnitude of a star, obtained by measuring the size of the star's image on an isochromatic photographic emulsion, using a filter transmitting only the longer wavelengths between 500 and 600 nanometers; nearly identical with visual magnitude. { ¦fōd·ō'vizh·ə·wəl 'mag·nə,tüd }

physical libration of the moon *See* lunar libration. { 'fiz·ə·kəl lī'brā·shən əv <u>th</u>ə 'mün }

Pictor [ASTRON] A southern constellation; right ascension 6 hours, declination 55°S. Also known as Easel. { 'pik,tór }

Pisces [ASTRON] A northern constellation; right ascension 1 hour, declination 15°N. Also know as Fishes. { 'pī·sēz }

Piscis Austrinus [ASTRON] A southern constellation; right ascension 22 hours, declination 30°S. Also known as Southern Fish. { 'pis·kəs 'ös·trī·nəs }

α **Piscis Austrinus** [ASTRON] The brightest star in the southern constellation Piscis Austrinus. Also known as Fomalhaut. { ¦al·fə 'pis·kəs 'ös·trī·nəs }

Piscis Volans *See* Volans. { 'pis·kəs 'vō·lənz }

plage [ASTRON] One of the luminous areas that appear in the vicinity of sunspots or disturbed areas on the sun; they may be seen distinctively in spectroheliograms taken in the calcium K line. { pläzh }

Planck era [ASTRON] The epoch in the early universe when the gravitational interaction between particles was as strong as the other interactions. { 'pläŋk ¸ir·ə }

planet [ASTRON] A relatively small celestial body moving in orbit around the sun or another star. { 'plan·ət }

planetarium [ASTRON] **1.** A projection device which accurately portrays the position of the stars and planets at any time in the past, present, or future from any point on the earth or the near region of space; the modern planetarium instrument is a mechanical-electrical analog of space. **2.** The name given to the building and gear associated with this device. { ¸plan·ə'ter·ē·əm }

planetary atmosphere [ASTRON] The outer shell of gas around some planets. { 'plan·ə¸ter·ē 'at·mə¸sfir }

planetary nebula [ASTRON] An oval or round nebula of expanding concentric rings of gas associated with a hot central star. { 'plan·ə¸ter·ē 'neb·yə·lə }

planetary nebula symbiotic *See* subdwarf symbiotic. { 'plan·ə¸ter·ē 'neb·yə·lə ¸sim·bē'äd·ik }

planetary orbit [ASTRON] The path that a planet has as it revolves about the sun. { 'plan·ə¸ter·ē 'ȯr·bət }

planetary perturbation [ASTRON] A deviation of a planet from its computed orbit because of the attraction of another celestial body or bodies. { 'plan·ə¸ter·ē ¸pər·tə'bā·shən }

planetary physics [ASTROPHYS] The study of the structure, composition, and physical and chemical properties of the planets of the solar system, including their atmospheres and immediate cosmic environment. { 'plan·ə¸ter·ē 'fiz·iks }

planetary precession [ASTRON] A comparatively small eastward motion of the equinoxes caused by the action of other planets in altering the plane of the earth's orbit. { 'plan·ə¸ter·ē pri'sesh·ən }

planetesimal [ASTRON] One of the rocky bodies, of the order of 1 mile (1.6 kilometer) in diameter, that are believed to have formed in the protosolar nebula, and whose accretion formed the rocky cores of the larger planets. { ¸plan·ə'tes·ə·məl }

planetocentric coordinates [ASTRON] Coordinates that indicate the position of a point on the surface of a planet, determined by the direction of a line joining the center of the planet to the point. { plə¦ned·ō¦sen·trik kō'ȯrd·ən·əts }

planetographic coordinates [ASTRON] Coordinates that indicate the position of a point on the surface of a planet, determined by the direction of a perpendicular to the mean surface at the point. { plə¦ned·ō¦graf·ik kō'ȯrd·ən·əts }

planetography [ASTRON] The descriptive science of the physical features of planets. { ¸plan·ə'täg·rə·fē }

planetoid *See* asteroid. { 'plan·ə¸tȯid }

planetology [ASTRON] Scientific study of the planets, in particular their surface markings. { ¸plan·ə'täl·ə·jē }

plasma cloud [ASTROPHYS] An aggregate of electrically charged particles that is embedded in the solar wind. { 'plaz·mə ¸klaud }

plasma engine [AERO ENG] An engine for space travel in which neutral plasma is accelerated and directed by external magnetic fields that interact with the magnetic field produced by current flow through the plasma. Also known as plasma jet. { 'plaz·mə ¸en·jən }

plasma propulsion [AERO ENG] Propulsion of spacecraft and other vehicles by using electric or magnetic fields to accelerate both positively and negatively charged particles (plasma) to a very high velocity. { 'plaz·mə prə'pəl·shən }

plasma rocket [AERO ENG] A rocket that is accelerated by means of a plasma engine. { 'plaz·mə ˌräk·ət }

plasma tail [ASTRON] A comet tail that is composed primarily of electrons and molecular ions, the dominant visible ion being positively ionized carbon monoxide, and that is generally straight, with a length in the range from 0.62 × 10^7 to 0.62 × 10^8 miles (1 × 10^7 to 1 × 10^8 kilometers). { 'plaz·mə 'tāl }

plateau ring structure [ASTRON] A lunar crater whose floor is significantly higher than the surrounding surface. { pla'tō 'riŋ ˌstrək·chər }

plate center [ASTRON] The point used as the origin of coordinates for measuring positions of stars on a photographic plate in photographic astrometry, ideally located on the optical axis of the telescope. { 'plāt ˌsen·tər }

plate constants [ASTRON] Coefficients that appear in linear equations used to derive the standard coordinates of the position of a star on a photographic plate from the measured coordinates of the star's image on the plate. { 'plāt ˌkän·stəns }

plate scale [ASTRON] The ratio of the angular distance between two stars to the linear distance between their images on a photographic plate. { 'plāt ˌskāl }

platonic year *See* great year. { plə'tän·ik 'yir }

Pleiades [ASTRON] An open cluster of a few hundred stars in the constellation Taurus; six of the stars are easily visible to the naked eye. { 'plē·ə,dēz }

plug nozzle [AERO ENG] A nozzle that is obtained by truncating a full-length spike nozzle, eliminating a significant portion of the spike nozzle without undue loss in propulsive thrust. { 'pləg ˌnäz·əl }

plunge *See* transit. { plənj }

plutino [ASTRON] A member of the Kuiper Belt that, like Pluto, is protected from close encounters with Neptune because its period is 3/2 Neptune's period. { plü'tē·nō }

Pluto [ASTRON] The most distant planet in the solar system; mean distance to the sun is about 3.7 × 10^9 miles (5.9 × 10^9 kilometers); it has no known satellite, and its sidereal revolution period is 248 years. { 'plüd·ō }

plutonic theory *See* volcanic theory. { plü'tän·ik 'thē·ə·rē }

pod [AERO ENG] An enclosure, housing, or detachable container of some kind on an airplane or space vehicle, as an engine pod. { päd }

Pogson scale [ASTRON] An index of brightness used in star catologs; it is the ratio of 2.512 to 1 between the brightness of successive magnitudes. { 'päg·sən ˌskāl }

Pointers [ASTRON] The stars α and β Ursae Majoris, which appear to point toward the north celestial pole and Polaris. { 'pȯint·ərz }

polar [ASTRON] A member of a class of cataclysmic variable stars whose light displays strong circular polarization. Also known as AM Herculis star. { 'pō·lər }

polar cap [ASTRON] Any of the bright areas covering the poles of Mars, believed to be composed of frozen carbon dioxide and water-ice. { 'pō·lər ˌkap }

polar distance [ASTRON] Angular distance from a celestial pole; the arc of an hour circle between a celestial pole, usually the elevated pole, and a point on the celestial sphere, measured from the celestial pole through 180°. { 'pō·lər 'dis·təns }

Polaris [ASTRON] A creamy supergiant star of stellar magnitude 2.0, spectral classification F8, in the constellation Ursa Minor; marks the north celestial pole, being about 1° from this point; the star Ursae Minoris. Also known as North Star; Pole Star. { pə'lar·əs }

polar night [ASTRON] The period of winter darkness in the polar regions, both northern and southern. { 'pō·lər 'nīt }

polar orbit [AERO ENG] A satellite orbit running north and south, so the satellite vehicle orbits over both the North Pole and the South Pole. { 'pō·lər 'ȯr·bət }

polar plumes [ASTRON] Columnlike plumes of hot coronal gas that are concentrated at the sun's magnetic poles. { 'pō·lər 'plümz }

polar sequence [ASTRON] A compilation of 96 brightness-standard stars within 2° of the North Pole. { 'pō·lər 'sē·kwəns }

Pole Star *See* Polaris.   { 'pōl ,stär }

Pollux   [ASTRON] A giant orange-yellow star with visual brightness of 1.16, a little less than 35 light-years from the sun, spectral classification K0-III, in the constellation Gemini; the star β Geminorum.   { 'päl·əks }

polygonal ring structure   [ASTRON] A lunar crater whose wall approximates a polygon in shape.   { pə'lig·ən·əl 'riŋ ,strək·chər }

poor cluster   [ASTRON] A galaxy cluster that has relatively few member galaxies.   { ¦pür ¦kləs·tər }

population I   [ASTRON] A class of stars which are relatively young, have relatively low peculiar velocities, and are found chiefly in the spiral arms of galaxies.   Also known as arm population.   { ,päp·yə'lā·shən 'wən }

population II   [ASTRON] A class of stars which are relatively old and evolved, have low metallic content and high peculiar velocities from 60 to 300 miles (100 to 500 kilometers) per second, and are found chiefly in the spheroidal halo of a galaxy. Also known as halo population.   { ,päp·yə'lā·shən 'tü }

population III   [ASTRON] A class of stars that condensed from the gas formed in the nucleosynthesis of the big bang, and consist entirely of hydrogen and helium.   { ,päp·yə'lā·shən 'thrē }

pore   [ASTRON] A very small, dark area on the sun formed by the separation of adjacent flocculi.   { pòr }

Portia   [ASTRON] A satellite of Uranus orbiting at a mean distance of 41,070 miles (66,100 kilometers) with a period of 12 hours 21 minutes, and with a diameter of about 68 miles (110 kilometers).   { 'pòr·shə }

positional astronomy   [ASTRON] The branch of astronomy that deals with the determination of the positions of celestial objects.   { pə'zish·ən·əl ə'strän·ə·mē }

position angle   [ASTRON] **1.** The angle formed by the great circle running through two celestial objects and the hour circle running through one of the objects. **2.** In measuring double stars, the angle formed between the great circle running through both components and the hour circle going through the primary measured from the north through the east from 0 to 360°.   { pə'zish·ən 'aŋ·gəl }

postnova   [ASTRON] A nova that has faded to the brightness it had before its outburst.   { pōs'nō·ə }

potentially hazardous asteroid   [ASTRON] An asteroid whose orbit approaches within 0.05 astronomical unit of the earth's orbit, and which is brighter than an absolute visual magnitude of 22.0, corresponding to a diameter of at least 110–240 meters (360–800 feet).   { pə,ten·chə·lē ,haz·ərd·əs 'as·tə,ròid }

Poynting-Robertson effect   [ASTRON] The gradual decrease in orbital velocity of a small particle such as a micrometeorite in orbit about the sun due to the absorption and reemission of radiant energy by the particle.   { 'pòint·iŋ 'räb·ərt·sən i,fekt }

practical astronomy   [ASTRON] That part of astronomy concerned with the use of information acquired by an observer in the solution of problems determining latitude and longitude on sea or land and directions on the earth's surface by the help of celestial objects.   { 'prak·ti·kəl ə'strän·ə·mē }

Praesepe   [ASTRON] A cluster of faint stars in the center of the constellation Cancer. Also known as Beehive; Manger.   { 'prē·sə,pē }

preatmospheric speed   [ASTRON] The speed of a meteoroid just before it enters the earth's atmosphere.   { ,prē,at·mə,sfir·ik 'spēd }

preceding limb   [ASTRON] The half of the limb of a celestial body with an observable disk that appears to precede the body in its apparent motion across the field of view of a fixed telescope.   { prē'sēd·iŋ 'lim }

precession in declination   [ASTRON] The component of general precession of the earth along a celestial meridian, amounting to about 20″ per year.   { prē'sesh·ən in ,dek·lə'nā·shən }

precession in right ascension   [ASTRON] The component of general precession of the earth along the celestial equator, amounting to about 46.1″ per year.   { prē'sesh·ən in 'rīt ə'sen·shən }

precession of nodes [ASTRON] The gradual change in direction of the orbital plane of a binary system. { prē'sesh·ən əv 'nōdz }

precession of the equinoxes [ASTRON] A slow conical motion of the earth's axis about the vertical to the plane of the ecliptic, having a period of 26,000 years, caused by the attractive force of the sun, moon, and other planets on the equatorial protuberance of the earth; it results in a gradual westward motion of the equinoxes. { prē'sesh·ən əv thə 'ē·kwə,näk·səz }

precomputed altitude [ASTRON] The altitude of a celestial body computed before observation with the sextant. Altitude corrections are included in the calculations but are applied with reversed sign. { 'prē·kəm,pyüd·əd 'al·tə,tüd }

preheating [ASTRON] A scenario suggested by detailed calculations of the inflationary universe cosmology in which the order parameter, as it relaxes about a new potential minimum following inflation, exhibits vibrations that can release energy in certain nonthermal modes. { prē'hēd·iŋ }

pre-Imbrian [ASTRON] Pertaining to the oldest lunar topographic features and lithologic map units constituting a system of rocks that appear in the mountainous terrae and are well displayed in the southern part of the visible lunar surface and over much of the reverse side. { prē'im·brē·ən }

prenova [ASTRON] A star that is destined to become a nova, but whose outburst has not yet taken place. { prē'nō·və }

pressure ionization [ASTROPHYS] A condition found in white dwarfs and other degenerate matter in which electron orbits overlap to the point that electrons in higher quantum levels are no longer associated with any particular nucleus and must be regarded as free. { 'presh·ər ,ī·ə·nə'zā·shən }

pressure suit [AERO ENG] A garment designed to provide pressure upon the body so that respiratory and circulatory functions may continue normally, or nearly so, under low-pressure conditions such as occur at high altitudes or in space without benefit of a pressurized cabin. { 'presh·ər ,süt }

pressure thrust [AERO ENG] In rocketry, the product of the cross-sectional area of the exhaust jet leaving the nozzle exit and the difference between the exhaust pressure and the ambient pressure. { 'presh·ər ,thrəst }

prestage [AERO ENG] A phase in the process of igniting a large liquid-fuel rocket in which the initial partial flow of propellants into the thrust chamber is ignited, and the combustion is satisfactorily established before the main stage is ignited. { 'prē,stāj }

primary [ASTRON] **1.** A planet with reference to its satellites, or the sun with reference to its planets. **2.** The brighter star of a double star system. { 'prī ,mer·ē }

primary body [ASTRON] The celestial body or central force field about which a satellite or other body orbits, or from which it is escaping, or toward which it is falling. { 'prī,mer·ē 'bäd·ē }

primordial black holes [ASTRON] Hypothetical black holes which may have formed in the early, highly compressed stages of the universe immediately following the big bang. { prī'mȯrd·ē·əl 'blak 'hōlz }

probe [AERO ENG] An instrumented vehicle moving through the upper atmosphere or space or landing upon another celestial body in order to obtain information about the specific environment. { prōb }

Procyon [ASTRON] A star of magnitude 0.3, of spectral type F5, and 11 light-years (1.04 × 10^{17} meters) from earth; one of a binary. Also known as α Canis Minoris. { 'prō·sē,än }

prograde motion [ASTRON] **1.** The apparent motion of a planet around the sun in the direction of the sun's rotation. **2.** See prograde orbit. { ¦prō'grād 'mō·shən }

prograde orbit [ASTRON] Orbital motion in the usual direction of celestial bodies within a given system; specifically, of a satellite, motion in the direction of rotation of the primary. Also known as prograde motion. { ¦prō'grād 'ȯr·bət }

program star [ASTRON] A star whose properties are observed or measured during a specified series of observations. { 'prō·grəm ,stär }

Prometheus [ASTRON] A satellite of Saturn which orbits at a mean distance of 139,000 kilometers (86,000 miles), just inside the F ring; together with Pandora, it holds this ring in place. { prə'mē·thē·əs }

prominence [ASTROPHYS] A volume of luminous, predominantly hydrogen gas that appears on the sun above the chromosphere; occurs only in the region of horizontal magnetic fields because these fields support the prominences against solar gravity. { 'präm·ə·nəns }

propellant injector [AERO ENG] A device for injecting propellants, which include fuel and oxidizer, into the combustion chamber of a rocket engine. { prə'pel·ənt in,jek·tər }

propellant mass ratio [AERO ENG] Of a rocket, the ratio of the effective propellant mass to the initial vehicle mass. Also known as propellant mass fraction. { prə'pel·ənt ¦mas ,rā·shō }

propellant weight fraction [AERO ENG] The weight of the solid propellant charge divided by weight of the complete solid propellant propulsion unit. { prə'pel·ənt ¦wāt ,frak·shən }

proper motion [ASTRON] That component of the space motion of a celestial body perpendicular to the line of sight, resulting in the change of a star's apparent position relative to that of other stars; expressed in angular units. { 'präp·ər 'mō·shən }

Prospector [AERO ENG] A specific uncrewed spacecraft designed to make a soft landing on the moon to take measurements, photographs, and soil samples, and then return to earth. { 'prä,spek·tər }

proplyd [ASTRON] A disk of dense gas and dust surrounding a young star. Derived from protoplanetary disk. { 'präp,lid }

Prospero [ASTRON] A small satellite of Uranus in a retrograde orbit with a mean distance of 10,250,000 miles (16,500,000 kilometers), eccentricity of 0.324, and sidereal period of 5.50 years. { 'präs·pə·rō }

Proteus [ASTRON] A satellite of Neptune orbiting at a mean distance of 73,100 miles (117,600 kilometers) with a period of 26.9 hours, and with a diameter of about 250 miles (400 kilometers). { 'prōd·ē·əs }

protogalaxy [ASTRON] A clump of matter in an early stage of galaxy formation in which the matter has started to collapse under its own gravity to form a galaxy. { ,prōd·ō'gal·ik·sē }

protoplanet [ASTRON] A precursor of one of the giant planets, which is believed to have formed, along with its satellites, from a minisolar nebula in a manner similar to that of the formation of the sun and planets. { 'prōd·ō,plan·ət }

protoplanetary disk See proplyd. { ,prōd·ō,plan·ä,ter·ē 'disk }

protostar [ASTRON] A dense condensation of material that is still in the process of accreting matter to form a star. { 'prōd·ə,stär }

protostele [ASTRON] A stele consisting of a solid rod of xylem surrounded by phloem { 'prōd·ə,stēl }

protosun [ASTRON] The condensation of material that lay at the center of the solar nebula and accreted material from it to form the sun. { 'prōd·ō,sən }

Proxima Centauri [ASTRON] The star that is the sun's nearest neighbor; stellar magnitude is 11, and it is 2° from the bright star α Centauri. { 'präk·sə·mə sen'tȯr·ē }

P spot [ASTRON] One of a pair of sunspots that appears to precede or lead the other across the face of the sun, or whose magnetic polarity is that which is normally found in such a sunspot during that sunspot cycle and in the hemisphere of the sun. { 'pē ,spät }

Psyche [ASTRON] An asteroid with a diameter of about 155 miles (249 kilometers), mean distance from the sun of 2.92 astronomical units, and unusual (M-type) surface composition; it may be made of solid metal. { 'sī·kē }

Ptolemaic system [ASTRON] The movements of the solar system according to Claudius Ptolemy; supposedly, the earth was a fixed center, with the sun and moon revolving about it in circular orbits; planets revolved in small circles (epicycles) whose centers revolved about the earth in larger circles (deferents). { ¦täl·ə¦mā·ik 'sis·təm }

Puck [ASTRON] A satellite of Uranus orbiting at a mean distance of 53,440 miles (86,010 kilometers) with a period of 18 hours 20 minutes, and with a diameter of about 96 miles (154 kilometers). { pək }

pulsar [ASTROPHYS] A celestial radio source, emitting intense short bursts of radio emission; the periods of known pulsars range between 33 milliseconds and 3.75 seconds, and pulse durations range from 2 to about 150 milliseconds with longer-period pulsars generally having a longer pulse duration. { 'pəl,sär }

pulsating star [ASTRON] Variable star whose luminosity fluctuates as the star expands and contracts; the variation in brightness is thought to come from the periodic change of radiant energy to gravitational energy and back. { 'pəl,sād·iŋ 'stär }

pulse window *See* mean profile. { 'pəls ,win·dō }

Puppis [ASTRON] A southern constellation; right ascension 8 hours, declination 40° south. Also known as Stern. { 'pəp·əs }

Puppis A [ASTRON] An extended, nonthermal radio source, the remnant of a supernova that exploded about 10,000 years ago, about 1.8 kiloparsecs (3.5 × 10^{16} miles or 5.6 × 10^{16} kilometers) from the earth. { 'pəp·əs 'ā }

pure Trojan group [ASTRON] The group of Trojan planets which lies near the Lagrangian point 60° behind Jupiter. Also known as Patroclus group. { 'pyúr 'trō·jən ,grüp }

Pyxis [ASTRON] A southern constellation; right ascension 9 hours, declination 30° south. Also known as Malus. { 'pik·səs }

Q

Q magnitude [ASTRON] The magnitude of a celestial object based on observations in the infrared at a wavelength of 19.5 micrometers. { 'kyü ,mag·nə,tüd }

QSO *See* quasar.

QSS *See* quasi-stellar radio source.

Quadrantids [ASTRON] A meteor shower whose radiant-right ascension of 15 hours and declination of +48° is in the constellation Boötes; velocity is 27 miles (43 kilometers) per second, and the strength is medium. { kwä'dran·tidz }

quadrature [ASTRON] The right-angle physical alignment of the sun, moon, and earth. { 'kwä·drə·chər }

quark star [ASTRON] A hypothetical star so dense that the nucleons have lost their identity and stability is derived from degenerate quarks. { 'kwärk ,stär }

quasar [ASTRON] Quasi-stellar astronomical object, often a radio source; all quasars have large red shifts; they have small optical diameter, but may have large radio diameter. Also known as quasi-stellar object (QSO). { 'kwā,zär }

quasi-geostationary satellites [AERO ENG] A constellation of satellites that simulates an object hovering over a particular location on the earth by having one member of the constellation moving slowly over a nearby location at all times. { ,kwäz·ē ,jē·ō,stā·shən·er·ē 'sad·əl,īts }

quasi-stellar object *See* quasar. { ¦kwä·zē 'stel·ər 'äb·jekt }

quasi-stellar radio source [ASTRON] A quasar that emits a significant fraction of its energy at radio frequencies ranging from 30 megahertz to 100 gigahertz. Abbreviated QSS. { ¦kwä·zē ¦stel·ər 'rād·ē·ō ,sórs }

quiescent prominence [ASTROPHYS] A vertical sheet of cool gas that is suspended in the solar corona for a period of days to months. { kwē'es·ənt 'präm·ə·nəns }

quiet sun [ASTROPHYS] The sun when it is free from unusual radio wave or thermal radiation such as that associated with sunspots. { 'kwī·ət 'sən }

quiet-sun noise [ASTROPHYS] Electromagnetic noise originating in the sun at a time when there is little or no sunspot activity. { 'kwī·ət 'sən ,nóiz }

R

R.A. *See* right ascension.

radar astronomy [ASTRON] The study of astronomical bodies and the earth's atmosphere by means of radar pulse techniques, including tracking of meteors and the reflection of radar pulses from the moon and the planets. { 'rā,där ə'strän·ə·mē }

radiant [ASTRON] **1.** A point on the celestial sphere through which pass the backward extensions of the trail of a meteor as observed at various locations, or the backward extensions of trails of a number of meteors traveling parallel to each other. **2.** A point on the celestial sphere toward which the stars in a moving cluster appear to travel. { 'rād·ē·ənt }

radiation-bounded nebula [ASTRON] An emission nebula whose central star is not hot enough to ionize the entire cloud. { ,rād·ē'ā·shən ¦baůn·dəd 'neb·yə·lə }

radiation era [ASTRON] The period in the early universe, lasting from roughly 20 seconds to 10^5 years after the big bang, when photons dominated the universe. { ,rād·ē'ā·shən ,ir·ə }

radiative braking [ASTRON] Deceleration of a star's rotation due to emission of electromagnetic radiation. { 'rād·ē,ād·iv 'brāk·iŋ }

radiative equilibrium [ASTROPHYS] The energy transfer through a star by radiation, absorption, and reradiation at a rate such that each section of the star is maintained at the appropriate temperature. { 'rād·ē,ād·iv ,ē·kwə'lib·rē·əm }

radio astronomy [ASTRON] The study of celestial objects by measurement and analysis of their emitted electromagnetic radiation in the wavelength range from roughly 1 millimeter to 30 millimeters. { 'rād·ē·ō ə'strän·ə·mē }

radio galaxy [ASTROPHYS] A galaxy that is emitting much energy in radio frequencies often from regions devoid of visible matter. { 'rād·ē·ō 'gal·ik·sē }

radio meteor [ASTRON] A meteor which has been detected by the reflection of a radio signal from the meteor trail of relatively high ion density (ion column); such an ion column is left behind a meteoroid when it reaches the region of the upper atmosphere between about 50 and 75 miles (80 and 120 kilometers), although occasionally radio meteors are detected at higher altitudes. { 'rād·ē·ō 'mēd·ē·ər }

radiometric magnitude [ASTRON] A celestial body's magnitude as calculated from the total amount of radiant energy of all the wavelengths that reach the earth's surface. { ¦rād·ē·ō¦me·trik 'mag·nə,tüd }

radio nebula [ASTROPHYS] A nebula that emits nonthermal radio-frequency radiation; derives its luminosity from collisions with the surrounding interstellar medium, or from processes associated with the magnetic fields presumably involved within the nebula; examples are the network nebulae in Cygnus and NGC 443. { 'rād·ē·ō 'neb·yə·lə }

radio source [ASTROPHYS] A source of extragalactic or interstellar electromagnetic emission in radio wavelengths. { 'rād·ē·ō ,sȯrs }

radio star [ASTROPHYS] A discrete celestial radio source. { 'rād·ē·ō ¦stär }

radio storm [ASTROPHYS] A prolonged period of disturbed emission or reception that lasts for periods of hours up to days. { 'rād·ē·ō ,stȯrm }

radio sun [ASTROPHYS] The sun as defined by its electromagnetic radiation in the radio portion of the spectrum. { 'rād·ē·ō ¦sən }

radio tail object [ASTROPHYS] An extragalactic object that displays a strong tail or jet at radio frequencies. { 'rād·ē·ō ¦tāl ¸äb·jəkt }

radius vector [ASTRON] A line joining the center of an orbiting body with the focus of its orbit located near its primary. { 'rād·ē·əs ¸vek·tər }

Ram *See* Aries. { ram }

ram pressure stripping [ASTRON] A process in which the gas in a galaxy interacts with the hot x-ray-emitting gas filling the cluster to which the galaxy belongs and is stripped away. { 'ram ¸presh·ər ¸strip·iŋ }

rapid proton capture process [ASTROPHYS] A mode of explosive nucleosynthesis in which each of the nuclei lighter than iron captures many protons, populating nuclides near the proton drip line, and these subsequently undergo a series of beta decays back to one of the stable nuclides. Also known as *rp*-process. { ¦rap·id ¸prō,tän 'kap·chər ¸prä,ses }

Ra-Shalom [ASTRON] An Aten asteroid that has a period of 0.759 year and an eccentricity of 0.436. { 'rä shə'lōm }

R association [ASTRON] A grouping of stars in a reflection nebula. { 'är ə,sō·sē,ā·shən }

rational horizon *See* celestial horizon. { 'rash·ən·əl hə'rīz·ən }

ray [ASTRON] One of the broad streaks that radiate from some craters on the moon, especially Copernicus and Tycho; they consist of material of high reflectivity and are seen from earth best at full moon. { rā }

ray crater [ASTRON] A large, relatively young lunar crater with visible rays. { 'rā ¸krād·ər }

ray system [ASTRON] The bright streaks radiating outward from a lunar crater. { 'rā ¸sis·təm }

R Coronae Borealis star [ASTROPHYS] A rare type of irregular variable star which has long periods of maximum brightness followed by a sudden, unpredictable reduction in brightness of several magnitudes, and a slower, sometimes erratic return to the original brightness. { ¦är kə¦rō,nē ¸bȯr·ē'al·əs ¸stär }

reactant ratio [AERO ENG] The ratio of the weight flow of oxidizer to fuel in a rocket engine. { rē'ak·tənt ¦rā·shō }

reaction propulsion [AERO ENG] Propulsion by means of reaction to a jet of gas or fluid projected rearward, as by a jet engine, rocket engine, or rocket motor. { rē'ak·shən prə,pəl·shən }

recession of galaxies [ASTROPHYS] The increase in the velocity of recession (red shift) of galaxies with distance from an observer on earth. { ri'sesh·ən əv 'gal·ik·sēz }

reconnection [ASTRON] The rejoining of solar magnetic field lines that have been severed at a neutral region. { ¸rē·kə'nek·shən }

recovery [AERO ENG] **1.** The procedure or action that obtains when the whole of a satellite, or a section, instrumentation package, or other part of a rocket vehicle, is retrieved after a launch. **2.** The conversion of kinetic energy to potential energy, such as in the deceleration of air in the duct of a ramjet engine. Also known as ram recovery. **3.** In flying, the action of a lifting vehicle returning to an equilibrium attitude after a nonequilibrium maneuver. { ri'kəv·ə·rē }

recovery area [AERO ENG] An area in which a satellite, satellite package, or spacecraft is recovered after reentry. { ri'kəv·ə·rē ¸er·ē·ə }

recovery capsule [AERO ENG] A space capsule designed to be recovered after reentry. { ri'kəv·ə·rē ¸kap·səl }

recovery package [AERO ENG] A package attached to a reentry or other body designed for recovery, containing devices intended to locate the body after impact. { ri'kəv·ə·rē ¸pak·ij }

recurrent nova [ASTRON] A binary star that undergoes outbursts every few decades in which the brightness increases roughly 100–1000 times, as the result of nuclear explosions in matter that has accreted on a white dwarf component star from a neighboring red giant component. { ri'kər·ənt 'nō·və }

recycle [AERO ENG] **1.** To stop the count in a countdown and to return to an earlier point. **2.** To give a rocket or other object a completely new checkout. { rē'sī·kəl }

red dwarf star [ASTRON] A red star of low luminosity, so designated by E. Hertzsprung; dwarf stars are commonly those main-sequence stars fainter than an absolute magnitude of +1, and red dwarfs are the faintest and coolest of the dwarfs. { 'red ¦dwȯrf 'stär }

red giant star [ASTRON] A star whose evolution has progressed to the point where hydrogen core burning has been completed, the helium core has become denser and hotter than originally, and the envelope has expanded to perhaps 100 times its initial size. { 'red ¦jī·ənt 'stär }

red giant tip [ASTRON] The upper tip of the red giant branch in the Hertzsprung-Russell diagram that represents stars undergoing a flash process. { 'red ¦jī·ənt 'tip }

redshift [ASTROPHYS] A systematic displacement toward longer wavelengths of lines in the spectra of distant galaxies and also of the continuous portion of the spectrum; increases with distance from the observer. Also known as Hubble effect. { 'red,shift }

Red Spot [ASTRON] A semipermanent marking of the planet Jupiter; some fluctuations in visibility exist; it does not rotate uniformly with the planet, indicating that it is a disturbance of Jupiter's atmosphere. { 'red ,spät }

reduced proper motion [ASTRON] The proper motion of a star expressed as a linear velocity. { ri'düst 'präp·ər 'mō·shən }

reduction of star places [ASTRON] The computation of mean places of stars from observations of their apparent places. { ri'dək·shən əv 'stär ,plās·əz }

reduction to the sun [ASTRON] In the spectroscopic determination of a star's radial motion referred to the sun, the correction that is needed to be applied to the observed radial velocity of the star to compensate for the motion of the earth with respect to the sun. { ri'dək·shən tü th̲ə 'sən }

red white-dwarf star [ASTRON] A star type that is considered an anomaly; these are objects 10,000 times fainter than the sun, with surface temperature below 4000 K so that surface radiation has cooled the star at an unexpectedly rapid rate. { 'red 'wīt ¦dwȯrf ,stär }

reentry [AERO ENG] The event when a spacecraft or other object comes back into the sensible atmosphere after being in space. { rē'en·trē }

reentry angle [AERO ENG] That angle of the reentry body trajectory and the sensible atmosphere at which the body reenters the atmosphere. { rē'en·trē ,aŋ·gəl }

reentry body [AERO ENG] That part of a space vehicle that reenters the atmosphere after flight above the sensible atmosphere. { rē'en·trē ,bäd·ē }

reentry nose cone [AERO ENG] A nose cone designed especially for reentry, consists of one or more chambers protected by a heat sink. { rē'en·trē 'nōz ,kōn }

reentry trajectory [AERO ENG] That part of a rocket's trajectory that begins at reentry and ends at the target or at the surface. { rē'en·trē trə,jek·trē }

reentry vehicle [AERO ENG] Any payload-carrying vehicle designed to leave the sensible atmosphere and then return through it to earth. { rē'en·trē ,vē·ə·kəl }

reentry window [AERO ENG] The area, at the limits of the earth's atmosphere, through which a spacecraft in a given trajectory can pass to accomplish a successful reentry for a landing in a desired region. { rē'en·trē ,win·dō }

reflection nebula [ASTRON] A type of bright diffuse nebula composed mainly of cosmic dust; it is visible because of starlight from nearby stars or nebula stars that is scattered by the dust particles. { ri'flek·shən ,neb·yə·lə }

reflector satellite [AERO ENG] Satellite so designed that radio or other waves bounce off its surface. { ri'flek·tər ,sad·əl,īt }

RE galaxy [ASTRON] A type of ring galaxy that consists of a single, relatively empty, ringlike structure, without any prominent condensation or nucleus. { ¦rē'ē ,gal·ik·sē }

regenerative engine [AERO ENG] **1.** A jet or rocket engine that utilizes the heat of combustion to preheat air or fuel entering the combustion chamber. **2.** Specifically, to a type of rocket engine in which one of the propellants is used to cool the engine by passing through a jacket prior to combustion. { rē'jen·rəd·iv 'en·jən }

regression of nodes [ASTRON] The westward movement of the nodes of the moon's orbit; one cycle is completed in about 18.6 years. { ri'gresh·ən əv 'nōdz }

regular cluster [ASTRON] A galaxy cluster that shows a smooth, centrally concentrated distribution of galaxies and an overall symmetric shape. { 'reg·yə·lər 'kləs·tər }

regular variable star [ASTRON] A variable star whose variation in brightness is repeated with a uniform period and light curve from one cycle to the next. { 'reg·yə·lər 'ver·ē·ə·bal 'stär }

Regulus [ASTRON] A star of stellar magnitude 1.34, about 67 light-years from the sun, spectral classification B8, in the constellation Leo; the star α Leonis. Also known as Little Ruler. { 'reg·yə·ləs }

reheating [ASTRON] A scenario suggested by the original calculations of the inflationary universe cosmology in which, following inflation, the universe quickly thermalizes to a temperature that is comparable to the energy density stored in the original symmetric pre-inflationary phase of matter. { rē'hēd·iŋ }

relative luminosity factor See luminosity function. { 'rel·əd·iv ,lü·mə'näs·əd·ē ,fak·tər }

relative orbit [ASTRON] The closed path described by the apparent position of the fainter member of a binary system relative to the brighter member. { 'rel·əd·iv 'òr·bət }

relative sunspot number [ASTRON] A measure of sunspot activity, computed from the formula $R = k(10g + f)$, where R is the relative sunspot number, f the number of individual spots, g the number of groups of spots, and k a factor that varies with the observer (his or her personal equation), the seeing, and the observatory (location and instrumentation). Also known as sunspot number; sunspot relative number; Wolf number; Wolf-Wolfer number; Zurich number. { 'rel·əd·iv 'sən,spät ,nəm·bər }

relativistic jet model [ASTRON] The accepted explanation of superluminal motion, wherein a feature moves from the nucleus of a quasar at relativistic speed and at a small angle to the line of sight to the earth. { ,rel·ə·tə,vis·tik 'jet ,mäd·əl }

rendezvous [AERO ENG] **1.** The event of two or more objects meeting with zero relative velocity at a preconceived time and place. **2.** The point in space at which such an event takes place, or is to take place. { 'rän·də,vü }

research rocket [AERO ENG] A rocket-propelled vehicle used to collect scientific data. { ri'sərch ,räk·ət }

réseau [ASTRON] A grid that is photographed by a separate exposure on the same plate as images of celestial objects. { rā'zō }

resistojet [AERO ENG] A type of propulsion system in which propellant is heated by passing it over resistively heated surfaces and then through a converging-diverging nozzle. { ri'zis·tə,jet }

resonance Kuiper Belt object [ASTRON] A member of the Kuiper Belt that lies in a stable region where it is protected from close encounters with the planet Neptune because its period of revolution is a simple fraction of Neptune's period. { ¦rez·ən·əns 'kī·pər ,belt ,äb,jekt }

resonant jet [AERO ENG] A pulsejet engine, exhibiting intensification of power under the rhythm of explosions and compression waves within the engine. { 'res·ən·ənt 'jet }

restart [AERO ENG] The act of firing a stage of a rocket after a previous powered flight. { 'rē,stärt }

restricted propellant [AERO ENG] A solid propellant having only a portion of its surface exposed for burning while the other surfaces are covered by an inhibitor. { ri'strik·təd prə'pel·ənt }

restricted proper motion [ASTRON] The rate of change of a star's apparent position relative to surrounding stars, corrected for precession, nutation, and aberration. { ri'strik·təd 'präp·ər 'mō·shən }

Reticulum [ASTRON] A southern constellation, right ascension 4 hours, declination 60° south. Also known as Net. { rə'tik·yə·ləm }

Reticulum system [ASTRON] A globular cluster or dwarf galaxy near the Large Magellanic Cloud. { ri'tik·yə·ləm ,sis·təm }

retrofire time [AERO ENG] The computed starting time and duration of firing of retro-rockets to decrease the speed of a recovery capsule and make it reenter the earth's atmosphere at the correct point for a planned landing. { 're·trō,fīr ,tīm }

retrograde motion [ASTRON] **1.** An apparent backward motion of a planet among the stars resulting from the observation of the planet from the planet earth which is also revolving about the sun at a different velocity. Also known as retrogression. **2.** See retrograde orbit. { 're·trə,grād 'mō·shən }

retrograde orbit [ASTRON] Motion in an orbit opposite to the usual orbital direction of celestial bodies within a given system; specifically, of a satellite, motion in a direction opposite to the direction of rotation of the primary. Also known as retro-grade motion. { 're·trə,grād 'òr·bət }

retrogression See retrograde motion. { ,re·trə'gresh·ən }

retrogressive metamorphism See retrograde metamorphism. { ¦re·trə¦gres·iv ,med·ə'mȯr,fiz·əm }

retrorocket [AERO ENG] A rocket fitted on or in a spacecraft, satellite, or the like to produce thrust opposed to forward motion. Also known as braking rocket. { ¦re·trō'räk·ət }

reversing layer [ASTROPHYS] A layer of relatively cool gas forming the lower part of the sun's chromosphere, just above the photosphere, that gives rise to absorption lines in the sun's spectrum. { ri'vərs·iŋ ,lā·ər }

R galaxy [ASTRON] A galaxy that displays rotational symmetry but lacks a clearly defined rotational or elliptical structure. { 'är ,gal·ik·sē }

RGU system [ASTRON] A system for obtaining a complete assessment of a star's magni-tude, based on measurements of the star's brightness when viewed through red, green, and ultraviolet filters. { ¦är¦jē'yü ,sis·təm }

Rhea [ASTRON] A satellite of Saturn, with estimated diameter of 450 miles (1530 kilome-ters). { 'rē·ə }

rice grains [ASTRON] Bright patches that stand out against the darker background of the surface of the sun; they are short-lived, and the pattern changes in a matter of minutes. { 'rīs ,grānz }

rich cluster [ASTRON] A galaxy cluster that has relatively many member galaxies. { ¦rich ¦kləs·tər }

Rigel [ASTRON] A multiple star of stellar magnitude 0.08, 650 light-years from the sun, spectral classification B8-Ia, in the constellation Orion; the star β Orionis. { rī·jəl }

right ascension [ASTRON] A celestial coordinate; the angular distance taken along the celestial equator from the vernal equinox eastward to the hour circle of a given celestial body. Abbreviated R.A. { 'rīt ə'sen·chən }

right sphere [ASTRON] The appearance of the celestial sphere as seen by an observer at the earth's equator. { 'rīt 'sfir }

rigidity [ASTROPHYS] The ratio of the momentum of a cosmic-ray particle to its electric charge, in units of the electron charge. { ri'jid·əd·ē }

Rigil Kent See Alpha Centauri. { 'rī·jəl 'kent }

Rigil Kentaurus See Alpha Centauri. { 'rī·jəl ken'tȯr·əs }

rill [ASTRON] A crooked, narrow crack on the moon's surface; may be a kilometer or more in width and a few to several hundred kilometers in length. Also spelled rille. { ril }

rill crater [ASTRON] A lunar crater that forms part of a rill. { 'ril ,krād·ər }

rille See rill. { ril }

ring A [ASTRON] The bright outer ring of Saturn, having an outside diameter of 169,000 miles (272,000 kilometers) and an inner diameter of 150,000 miles (242,000 kilome-ters). { 'riŋ 'ā }

ring B [ASTRON] The brightest of Saturn's rings, with an outer diameter of 146,000 miles (235,000 kilometers) and an inner diameter of 114,000 miles (183,000 kilome-ters). { 'riŋ 'bē }

ring C [ASTRON] A faint ring of Saturn inside ring B having an outer diameter of 114,000 miles (183,000 kilometers) and an inner diameter of 91,000 miles (146,000 kilometers). Also known as crepe ring. { 'riŋ 'sē }

ring D |ASTRON| A very faint ring of Saturn that is fainter than ring C and lies between ring C and the planet's surface. { 'riŋ 'dē }

ring E |ASTRON| A diffuse, faint outer ring of Saturn that extends from inside the orbit of Enceladus at about 112,000 miles (181,000 kilometers) from Saturn to outside the orbit of Dione at about 300,000 miles (480,000 kilometers) from the planet. { 'riŋ 'ē }

ring F |ASTRON| A narrow ring of Saturn just outside ring A that consists of more than five separate strands and is held in place by two small satellites. { 'riŋ 'ef }

ring G |ASTRON| A tenuous ring of Saturn with an average diameter of 211,000 miles (340,000 kilometers), and lying 21,000 miles (34,000 kilometers) outside ring A. { 'riŋ 'jē }

ring galaxy |ASTRON| A class of galaxy whose ringlike structure has clumps of ionized hydrogen clouds on its periphery, may have a nucleus of stars, and is usually accompanied by a small galaxy; probably formed when a small galaxy crashes through the disk of a spiral galaxy. { 'riŋ ‚gal·ik·sē }

Ring Nebula |ASTRON| A nebula in the summer constellation Lyra; it is an example of the planetary type of gaseous nebulae. { 'riŋ 'neb·yə·lə }

ring plain |ASTRON| A lunar crater of exceptionally large diameter and with a relatively smooth interior. { 'riŋ ‚plān }

rings of Saturn |ASTRON| Circular rings that encircle the planet Saturn at its equator; there are four main regions to the ring system; theory and observations indicate that the rings are composed of separate particles which move independently in the four series of circular coplanar orbits. { 'riŋz əv 'säd·ərn }

ringwall structure |ASTRON| A lunar crater whose center lies on the wall, or on the line of the wall, of a larger crater. { 'riŋ‚wȯl ‚strək·chər }

rise |ASTRON| Of a celestial body, to cross the visible horizon while ascending. { rīz }

River Po See Eridanus. { 'riv·ər 'pō }

RK galaxy |ASTRON| A type of ring galaxy which consists of a ringlike structure with a large, bright condensation or knot within the ring itself. { ¦är'kā ‚gal·ik·sē }

R magnitude |ASTRON| The wavelength of a celestial object based on observations at a wavelength of 680 nanometers. { 'är ‚mag·nə‚tüd }

R Monocerotis |ASTRON| An irregular variable star at the tip of the small fan-shaped emission nebula NGC 2261. { 'är ‚män·ə'ser·əd·əs }

RN galaxy |ASTRON| A type of ring galaxy which consists of a ringlike structure with a nucleus somewhere within it, the nucleus being somewhat like those seen in ordinary spiral galaxies but typically lying off the center of the ring. { ¦är'en ‚gal·ik·sē }

Roche's limit |ASTROPHYS| The limiting distance below which a satellite orbiting a celestial body would be disrupted by the tidal forces generated by the gravitational attraction of the primary; the distance depends on the relative densities of the bodies and on the orbit of the satellite; it is computed by $R = 2.45(Lr)$, where L is a factor that depends on the relative densities of the satellite and the body, R is the radius of the satellite's orbit measured from the center of the primary body, and r is the radius of the primary body; if the satellite and the body have the same density, the relationship is $R = 2.45r$. { 'rō·shəz ‚lim·ət }

rocket |AERO ENG| **1.** Any kind of jet propulsion capable of operating independently of the atmosphere. **2.** A complete vehicle driven by such a propulsive system. { 'räk·ət }

rocket astronomy |ASTRON| The discipline comprising measurements of the electromagnetic radiation from the sun, planets, stars, and other bodies, of wavelengths that are almost completely absorbed below the 150-mile (250-kilometer) level, by using a rocket to carry instruments above 150 miles to measure these phenomena. { 'räk·ət ə‚strän·ə·mē }

rocket chamber |AERO ENG| A chamber for the combustion of fuel in a rocket; in particular, that section of the rocket engine in which combustion of propellants takes place. { 'räk·ət ‚chām·bər }

rocket engine [AERO ENG] A reaction engine that contains within itself, or carries along with itself, all the substances necessary for its operation or for the consumption or combustion of its fuel, not requiring intake of any outside substance and hence capable of operation in outer space. Also known as rocket motor. { 'räk·ət ˌen·jən }

rocket igniter [AERO ENG] An igniter for the propellant in a rocket. { 'räk·ət igˌnīd·ər }

rocket propulsion [AERO ENG] Reaction propulsion by a rocket engine. { 'räk·ət proˌpəl·shən }

rocketry [AERO ENG] **1.** The science or study of rockets, embracing theory, research, development, and experimentation. **2.** The art and science of using rockets, especially rocket ammunition. { 'räk·ə·trē }

rocket staging [AERO ENG] The use of successive rocket sections or stages, each having its own engine or engines; each stage is a complete rocket vehicle in itself. { 'räk·ət ˌstāj·iŋ }

rocket thrust [AERO ENG] The thrust of a rocket engine. { 'räk·ət ˌthrəst }

Rogallo wing [AERO ENG] A glider folded inside a spacecraft; to be deployed during the spacecraft's reentry like a parachute, gliding the spacecraft to a landing. { rō'gäl·ō ˌwiŋ }

Rood-Sastry classification [ASTRON] A classification scheme for galaxy clusters that differentiates between a number of basic cluster morphologies. { ˈrüd ˈsas·trē ˌklas·ə·fəˈkā·shən }

Rosalind [ASTRON] A satellite of Uranus orbiting at a mean distance of 43,450 miles (69,930 kilometers) with a period of 13 hours 26 minutes, and with a diameter of about 36 miles (58 kilometers). { 'räz·lind }

Rosette Nebula [ASTRON] A nebula classified as NGC 2237; this nebula contains numerous small dense clouds that have been photographed with large telescopes. { rō'zet 'neb·yə·lə }

rough burning [AERO ENG] Pressure fluctuations frequently observed at the onset of burning and at the combustion limits of a ramjet or rocket. { 'rəf 'bərn·iŋ }

rp-process See rapid proton capture process. { ˈärˈpē 'präˌses }

RR Lyrae stars [ASTRON] Pulsating variable stars with a period of 0.05–1.2 days in the halo population of the Milky Way Galaxy; color is white, and they are mostly stars of spectral class A. Also known as cluster cepheids; cluster variables. { ˈärˈär 'lī·rē ˌstärz }

RS Canum Venaticorum stars [ASTRON] A group of peculiar binary stars with orbital periods between 1 and 14 days, in which the cores of the H and K lines of singly ionized calcium display strong emission, and the hotter of the two stars is of spectral type F or G. { ˈärˈes ˈkan·ən vəˌnad·əˈkór·əm ˌstärz }

R star [ASTRON] A star of spectral type R, having spectral characteristics similar to those of types G and K, except that molecular bands of molecular carbon (C_2), cyanogen radical (CN), and methyldadyne (CH) are prominent. { 'är ˌstär }

Rule See Norma. { 'rül }

runaway star [ASTRON] A star of spectral type O or early B with an unusually high spatial velocity; believed to result from a supernova explosion in a close binary system. { 'rən·ə'wā 'stär }

Russell diagram See Hertzsprung-Russell diagram. { 'rəs·əl ˌdī·əˌgram }

Russell mixture [ASTROPHYS] A mixture of elements with the same relative proportions as are found in the sun and other stars. { 'rəs·əl 'miks·chər }

Russell-Vogt theorem See Vogt-Russell theorem. { 'rəs·əl 'vót 'thir·əm }

RV Tauri stars [ASTRON] A class of stars; they are long-period pulsating variable types with periods from about 50 to 150 days; otherwise they are like the shorter-period W Virginis stars; they are found in both the Milky Way Galaxy and the globular clusters. { ˈärˈvē 'tór·ē ˌstärz }

RW Aurigae stars [ASTRON] A class of stars that are variable, and whose light variations are rapid and irregular. { ˈärˈdəb·əlˌyü ó'rī·gē ˌstärz }

S

Sagitta [ASTRON] A small constellation; right ascension 20 hours, declination 10° north. Also known as Arrow. { sə'jid·ə }

Sagittarius [ASTRON] A constellation whose major portion lies in the Milky Way; right ascension 19 hours, declination 25° south. Also known as Archer. { ‚saj·ə'ter·ē·əs }

Sagittarius A [ASTRON] An intense radio source in the constellation Sagittarius, apparently comprising a gaseous envelope surrounding a small dense core that is believed to constitute the center of the Milky Way Galaxy. { ‚saj·ə'ter·ē·əs 'ā }

Sagittarius arm [ASTRON] A spiral arm of the Milky Way Galaxy that lies between the sun and the galactic center in the direction of the constellation Sagittarius. { ‚saj·ə'ter·ē·əs 'ärm }

Sagittarius B2 [ASTRON] The richest molecular radio source in the Galaxy, located near the galactic center and consisting of a massive, dense complex of at least seven HII regions and molecular clouds. { ‚saj·ə'ter·ē·əs ¦bē'tü }

Sagittarius star cloud [ASTRON] A large star cloud within the Milky Way; its extension is about 1500 to 6000 light-years (1.42×10^{19} to 5.68×10^{19} meters) from the sun. { ‚saj·ə'ter·ē·əs 'stär ‚klaůd }

Sail See Vela. { sāl }

salpeter process See three-alpha process. { sal'pēd·ər ‚präs·əs }

saros [ASTRON] A cycle of time after which the centers of the sun and moon, and the nodes of the moon's orbit return to the same relative position; this period is 18 years $11^{1}/_{3}$ days, or 18 years $10^{1}/_{3}$ days if 5 rather than 4 leap years are included. { 'sa‚räs }

Sa spiral [ASTRON] A class of spiral galaxy, including those galaxies that have the largest center sections and closely wound galactic arms. { ¦es¦ā 'spī·rəl }

satellite [AERO ENG] See artificial satellite. [ASTRON] A small, solid body moving in an orbit around a planet; the moon is a satellite of earth. { 'sad·əl‚īt }

satellite astronomy [ASTRON] The study of astronomical objects by using detectors mounted on earth-orbiting satellites or deep-space probes; allows observations that are not obstructed by the earth's atmosphere. { 'sad·əl‚īt ə'strän·ə·mē }

satellite tracking [AERO ENG] Determination of the positions and velocities of satellites through radio and optical means. { 'sad·əl‚īt ‚trak·iŋ }

satelloid [AERO ENG] A vehicle that revolves about the earth or other celestial body, but at such altitudes as to require sustaining thrust to balance drag. { 'sad·əl‚ȯid }

Saturn [AERO ENG] One of the very large launch vehicles built primarily for the Apollo program; begun by Army Ordnance but turned over to the National Aeronautics and Space Administration for the manned space flight program to the moon. [ASTRON] The second largest planet in the solar system (mass is 95.2 compared to earth's 1) and the sixth in the order of distance to the sun; it is visible to the naked eye as a yellowish first-magnitude star except during short periods near its conjunction with the sun; it is surrounded by a series of rings. { 'sad·ərn }

saturnicentric coordinates [ASTRON] Coordinates that indicate the position of a point on the surface of Saturn, determined by the direction of a line joining the center of Saturn to the point. { sə¦tər·nə¦sen·trik kō'ȯrd·ən‚ats }

saturnigraphic coordinates [ASTRON] Coordinates that indicate the position of a point on the surface of Saturn, determined by the direction of a line perpendicular to the mean surface at the point. { sə¦tər·nə¦graf·ik kō'ȯrd·ən‚ats }

Saturn Nebula

Saturn Nebula [ASTRON] A double-ring planetary nebula in the constellation Aquarius, about 700 parsecs away. { 'sa,tərn ,neb·yə·lə }

saucer crater [ASTRON] A very shallow type of bowl crater on the moon. { 'sȯs·ər ,krād·ər }

Sb spiral galaxy [ASTRON] A class of spiral galaxy characterized by smaller central bodies and more open, larger arms. { ¦es,bē 'spī·rəl 'gal·ik·sē }

scattered Kuiper Belt object [ASTRON] A member of the Kuiper belt with a very large, very elliptical orbit, whereby its greatest distance from the sun can range up to 200 astronomical units or more. { ¦skad·ərd 'kī·pər,belt ,äb,jekt }

Scheat [ASTRON] A red giant, irregular, variable star, in the constellation Pegasus. { shē'at }

Schönberg-Chandrasekhar limit *See* Chandrasekhar-Schönberg limit. { 'shərn,bȯrg ¦chan·drə'sä·kər ,lim·ət }

Schroter effect [ASTRON] The occurrence of the dichotomy of Venus earlier than theoretically predicted when it is waning, and later than theoretically predicted when waxing. { 'shrōd·ər i,fekt }

Schwarzschild criterion [ASTROPHYS] A criterion for determining the stability of a stellar medium against convective motion, according to which convection takes place when the temperature gradient is greater than the gradient that would exist if the medium were adiabatic. { 'shvärts,shilt krī,tir·ē·ən }

Schwassman-Wachmann comet [ASTRON] A variable cometlike asteroid whose period is 16 years; its orbit is very nearly circular and lies between the orbits of Saturn and Jupiter. { 'shväs,män 'vak,män ,käm·ət }

Sco-Cen association [ASTRON] An association of very young stars within the Gould belt whose brightest member is Antares. { 'skō 'sen ə,sō·sē,ā·shən }

Scorpion *See* Scorpius. { 'skȯr·pē·ən }

Scorpio X-1 [ASTROPHYS] The most intense celestial source of x-rays known, associated with a highly variable radio source and a variable optical source, ranging from the twelfth to fourteenth magnitude. Abbreviated Sco X-1. { 'skȯr·pē·ō ¦eks'wən }

Scorpius [ASTRON] A southern constellation, right ascension 16 hours, declination 40° south; the bright-red star Antares is located in it. Also known as Scorpion. { 'skȯr· pē·əs }

Scout [AERO ENG] A four-stage all-solid-propellant rocket, used as a space probe and orbital test vehicle; first launched July 1, 1960, with a 150-pound (68-kilogram) payload. { skaút }

Sco X-1 *See* Scorpio X-1.

screaming [AERO ENG] A form of combustion instability, especially in a liquid-propellant rocket engine, of relatively high frequency, characterized by a high-pitched noise. { 'skrēm·iŋ }

scrub [AERO ENG] To cancel a scheduled firing, either before or during countdown. { skrəb }

Sc spiral galaxy [ASTRON] A class of spiral galaxy characterized by spirals with the largest and most loosely coiled arms and the smallest central portion. { ¦es¦sē ¦spī· rəl 'gal·ik·sē }

SC star [ASTRON] A star of a type intermediate between carbon stars and S stars. { ¦es¦sē ,stär }

Sculptor [ASTRON] A southern constellation, right ascension 0 hours, declination 30° south. Also known as Sculptor's Apparatus; Workshop. { 'skəlp·tər }

Sculptor Group [ASTRON] One of the nearest groups of galaxies beyond the Local Group, consisting of about five large galaxies near the south galactic pole. { 'skəlp· tər ,grüp }

Sculptor's Apparatus *See* Sculptor. { 'skəlp·tərz ,ap·ə'rad·əs }

Sculptor system [ASTRON] A dwarf, elliptical galaxy in the Local Group, about 270,000 light years away. { 'skəlp·tər ,sis·təm }

Scutum [ASTRON] A southern constellation, right ascension 19 hours, declination 10° south. Also known as Shield. { 'sküd·əm }

Sea Goat *See* Capricornus. { 'sē ,gōt }

sealed cabin [AERO ENG] The occupied space of an aircraft or spacecraft characterized by walls which do not allow gaseous exchange between the inner atmosphere and its surrounding atmosphere, and containing its own mechanisms for maintenance of the inside atmosphere. { 'sēld 'kab·ən }

search for extraterrestrial intelligence [ASTRON] A systematic effort to discover evidence for the existence beyond the earth of other advanced civilizations. Abbeviated SETI. { ¦sərch fər ¸ek·strə·tə¦res·trē·əl in'tel·ə·jəns }

sea rim [ASTRON] The apparent horizon as actually observed at sea. { 'sē ¸rim }

sector boundary [ASTROPHYS] The rapid transition from one polarity to another in the interplanetary magnetic field. { 'sek·tər ¸baún·drē }

sector structure [ASTROPHYS] The polarity pattern of the interplanetary magnetic field observed during a solar rotation. { 'sek·tər ¸strək·chər }

secular acceleration [ASTRON] An apparent gradual acceleration of the moon's motion in its orbit, as measured relative to mean solar time. { 'sek·yə·lər ak¸sel·ə'rā·shən }

secular parallax [ASTRON] An apparent angular displacement of a star, resulting from the sun's motion. { 'sek·yə·lər 'par·ə¸laks }

secular perturbations [ASTROPHYS] Changes in the orbit of a planet, or of a satellite, that operates in extremely long cycles. { 'sek·yə·lər ¸pər·dər'bā·shənz }

secular variable [ASTRON] A star whose brightness appears to have slowly lessened or increased over a time period of centuries. { 'sek·yə·lər 'ver·ē·ə·bəl }

secular variation [ASTRON] A perturbation of the moon's motion caused by variations in the effect of the sun's gravitational attraction on the earth and moon as their relative distances from the sun vary during the synodic month. { 'sek·yə·lər ¸ver·ē'ā·shən }

seed nuclei [ASTROPHYS] Nuclei from which other nuclei are synthesized in stars. { 'sēd ¸nü·klē¸ī }

seeing [ASTRON] The clarity and steadiness of an image of a star in a telescope. { 'sē·iŋ }

selected areas *See* Kapetyn selected areas. { si'lek·təd 'er·ē·əz }

selenocentric [ASTRON] Pertaining to the moon's center. { sə¦lē·nō'sen·trik }

selenodesy [ASTRON] The branch of applied mathematics that determines, by observation and measurement, the exact positions of points on the moon's surface, as well as the shape and size of the moon. { sə¦lē·nə¦des·ē }

selenodetic [ASTRON] Of, or pertaining to, or determined by selenodesy. { sə¦lē·nə¦ded·ik }

selenofault [ASTRON] A geological fault in the lunar surface. { sə'lē·nə¸fȯlt }

selenographic coordinates [ASTRON] A coordinate system for specifying positions on the moon's surface relative to the moon's center, consisting of selenographic latitude and longitude, or of a cartesian coordinate system. { sə¦lē·nə¦graf·ik kō'ȯrd·ən·əts }

selenographic latitude [ASTRON] The angular distance, measured along a meridian, between a point on the moon's surface and the moon's equator. { sə¦lē·nə¦graf·ik 'lad·ə¸tüd }

selenographic longitude [ASTRON] The angular distance, measured along the moon's equator, between the meridian passing through a point on the moon's surface and the first lunar meridian. { sə¦lē·nə¦graf·ik 'län·jə¸tüd }

selenography [ASTRON] Studies pertaining to the physical geography of the moon; specifically, referring to positions on the moon measured in latitude from the moon's equator and in longitude from a reference meridian. { ¸sel·ə·näg·rə·fē }

selenology [ASTRON] A branch of astronomy that treats of the moon, including such attributes as magnitude, motion and constitution. Also known as lunar geology. { ¸sel·ə'näl·ə·jē }

selenomorphology [ASTRON] The study of landforms on the moon, including their origin, evolution, and distribution. { sə'lē·nō·mȯr'fäl·ə·jē }

semidetached binary [ASTRON] A binary system whose secondary member fills its Roche lobe but whose primary member does not. { ¦sem·i·di'tacht 'bī¸ner·ē }

semidiameter [ASTRON] Measured at the observer, half the angle subtended by the visible disk of a celestial body. { ¦sem·i·dī'am·əd·ər }

semidiurnal [ASTRON] Having a period of, occurring in, or related to approximately half a day. { ¦sem·i·dī'ərn·əl }

semiregular variables [ASTRON] Variable red giant stars whose variation in brightness is repeated, but whose period and light curve may vary considerably from one cycle to the next; they have absolute magnitude of about 0 or −1 and quasi-periods of from about 40 to 150 days. { ¦sem·i'reg·yə·lər 'ver·ē·ə·bəlz }

semisynchronous satellite [AERO ENG] An artificial earth satellite that makes one revolution in exactly one-half of a sidereal day (11 hours 58 minutes 2 seconds). { ¦sem·ē¸siŋ·krə·nəs 'sad·əl¸īt }

sensible horizon [ASTRON] That circle of the celestial sphere formed by the intersection of the celestial sphere and a plane through any point, such as the eye of an observer, and perpendicular to the zenith-nadir line. { 'sen·sə·bəl hə'rīz·ən }

separation [AERO ENG] The action of a fallaway section or companion body as it casts off from the remaining body of a vehicle, or the action of the remaining body as it leaves a fallaway section behind it. { ¸sep·ə'rā·shən }

Serpens [ASTRON] A constellation, right ascension 17 hours, declination 0°. Also known as Serpent. { 'sər¸penz }

Serpent See Serpens. { 'sər·pənt }

Serpent Bearer See Ophiucus. { 'sər·pənt 'ber·ər }

set [ASTRON] Of a celestial body, to cross the visible western horizon while descending. { set }

Setebos [ASTRON] A small satellite of Uranus in a retrograde orbit with a mean distance of 11,090,000 miles (17,850,000 kilometers), eccentricity of 0.522, and sidereal period of 6.25 years. { 'sed·ə¸bōs }

SETI See search for extraterrestrial intelligence. { 'sed·ē }

Sextans [ASTRON] A constellation in the southern hemisphere, right ascension 10 hours, declination 0°. Also known as Sextant. { 'sek¸stanz }

Sextant See Sextans. { 'sek·stənt }

sextile aspect [ASTRON] The position of two celestial bodies when they are 60° apart. { 'sek¸stīl 'as¸pekt }

Seyfert galaxy [ASTRON] A galaxy that has a small, bright nucleus from which violent explosions may occur. { 'zī·fərt ¸gal·ik·sē }

Seyfert's Sextet [ASTRON] A compact collection of galaxies that surrounds the galaxy NGC 6027 and has both spiral and irregular members, most of which interact with each other. { 'zī·fərts seks'tet }

shadow bands [ASTRON] Rippling bands of shadow that appear on every white surface of flat terrestrial objects a few minutes before the total eclipse of the sun. { 'shad·ō ¸banz }

shadow-transit [ASTRON] A transit of one of Jupiter's Galilean satellites, in which the satellite casts its shadow on the planet's disk. { ¦shad·ō 'tranz·ət }

Shahbazian objects [ASTRON] Compact collections of the order of 10 galaxies. { shə'bä·zē·ən ¸äb·jeks }

Shaula [ASTRON] A blue-white subgiant star of stellar magnitude 1.7, spectral classification B2-IV, in the constellation Scorpius; the star λ Scorpii. { 'shaủ·lə }

shell star [ASTRON] A type of star which is believed to be surrounded by a tenuous envelope of gas, as indicated by bright emission lines in its spectrum. { 'shel ¸stär }

shepherding satellite [ASTRON] A satellite that helps to hold in place a given ring of a planet. { ¦shep·ərd·iŋ 'sad·əl¸īt }

Shield See Scutum. { shēld }

Ship See Argo. { ship }

shooting star [ASTRON] A small meteor that has the brief appearance of a darting, starlike object. { 'shüd·iŋ 'stär }

short-period comet [ASTRON] A comet whose period is short enough for observations at two or more apparitions to be interrelated; usually taken to be a comet whose period is shorter than 200 years. { 'shȯrt ¦pir·ē·əd 'käm·ət }

shot [AERO ENG] An act or instance of firing a rocket, especially from the earth's surface. { shät }

shower meteor [ASTRON] A meteor whose direction of arrival is approximately parallel to others belonging to the same meteor shower. { 'shaů·ər ¸mēd·ē·ər }

shutoff [AERO ENG] In rocket propulsion, the intentional termination of burning by command from the ground or from a self-contained guidance system. { 'shət¸óf }

Sickle [ASTRON] A group of six stars in the constellation Leo that outline the head of the lion. { 'sik·əl }

sidereal [ASTRON] Referring to a quantity, such as time, to indicate that it is measured in relation to the apparent motion or position of the stars. { sī'dir·ē·əl }

sidereal day [ASTRON] The time between two successive upper transits of the vernal equinox; this period measures one sidereal day. { sī'dir·ē·əl 'dā }

sidereal hour angle [ASTRON] The angle along the celestial equator formed between the hour circle of a celestial body and the hour circle of the vernal equinox, measuring westward from the vernal equinox through 360°. { sī'dir·ē·əl ¦aůr ¸aŋ·gəl }

sidereal month [ASTRON] The time period of one revolution of the moon about the earth relative to the stars; this period varies because of perturbations, but it is a little less than $27\frac{1}{3}$ days. { sī'dir·ē·əl 'mənth }

sidereal noon [ASTRON] The instant in time that the vernal equinox is on the meridian. { sī'dir·ē·əl 'nün }

sidereal period [ASTRON] The length of time required for one revolution of a celestial body about its primary, with respect to the stars. { sī'dir·ē·əl 'pir·ē·əd }

sidereal time [ASTRON] Time based on diurnal motion of stars; it is used by astronomers but is not convenient for ordinary purposes. { sī'dir·ē·əl 'tīm }

sidereal year [ASTRON] The time period relative to the stars of one revolution of the earth around the sun; it is about 365.2564 mean solar days. { sī'dir·ē·əl 'yir }

sign of the zodiac [ASTRON] The zodiac is divided into 12 sections, called signs, in each of which the sun is situated for 1 month of the year; each sign, 30° in length, is named from a constellation with which the sign once coincided. { 'sīn əv thə 'zō·dē¸ak }

simple ring See elementary ring structure. { 'sim·pəl 'riŋ }

simultaneous altitudes [ASTRON] Altitudes of two or more celestial bodies observed at the same time. { ¸sī·məl'tā·nē·əs 'al·tə¸tüdz }

single-stage rocket [AERO ENG] A rocket or rocket missile to which the total thrust is imparted in a single phase, by either a single or multiple thrust unit. { 'siŋ·gəl ¦stāj 'räk·ət }

Sinope [ASTRON] A small satellite of Jupiter with a diameter of about 17 miles (27 kilometers), orbiting with retrograde motion at a mean distance of about 1.47×10^7 miles (2.37×10^7 kilometers). Also known as Jupiter IX. { 'sin·ə·pē }

siriometer [ASTRON] A unit of length, formerly used in astronomical measurement, equal to 10^6 astronomical units, or 1.496×10^{17} meters. { ¸sir·ē'äm·əd·ər }

Sirius [ASTRON] The brightest-appearing star in the sky; 8.7 light-years from the sun, spectral class A1V; it has a white dwarf companion. Also known as Dog Star. { 'sir·ē·əs }

skip vehicle [AERO ENG] A reentry body which climbs after striking the sensible atmosphere in order to cool the body and to increase its range. { 'skip ¸vē·ə·kəl }

sky [ASTRON] In the daytime the apparent blue dome resting on the earth along the horizon circle; at night the blue becomes nearly black. { skī }

sky diagram [ASTRON] A diagram of the heavens, indicating the apparent positions of various celestial bodies with reference to the horizon system of coordinates. { 'skī ¸dī·ə¸gram }

skylight See diffuse sky radiation. { 'skī¸līt }

sky map [ASTRON] A planar representation of areas of the sky showing positions of celestial bodies. { 'skī ¸map }

sky radiation See diffuse sky radiation. { 'skī ¸rād·ē¸ā·shən }

slow nova [ASTRON] A nova whose brightness takes a month or more to reach a maximum, and many years to decrease to the original value. { 'slō 'nō·və }

Small Magellanic Cloud [ASTRON] The smaller of the two star clouds near the south celestial pole; it is about 170,000 light-years away and contains a wide assortment of giant and variable stars, star clusters, and nebulae. Also known as Nubecula Minor. { 'smȯl ¦maj·ə¦lan·ik 'klaüd }

SMC X-1 [ASTRON] The most luminous x-ray pulsar known, located in the small Magellanic Cloud.

SNR *See* supernova remnant.

SNU *See* solar neutrino unit.

soft landing [AERO ENG] The act of landing on the surface of a planet or moon without damage to any portion of the vehicle or payload, except possibly the landing gear. { 'sȯft 'land·iŋ }

Sol *See* sun. { säl }

solar activity [ASTRON] Disturbances on the surface of the sun; examples are sunspots, prominences, and solar flares. { 'sō·lər ak'tiv·əd·ē }

solar antapex [ASTRON] The point on the celestial sphere away from which the solar system is moving; it lies in the constellation Columba. { 'sō·lər ant'ā,peks }

solar apex [ASTRON] A point toward which the solar system is moving; it is about 10° southwest of the star Vega. { 'sō·lər 'ā,peks }

solar bridge [ASTRON] A bright, narrow, streak-shaped region which is sometimes observed across a large sunspot, dividing the umbra into two or more parts. { 'sō·lər 'brij }

solar burst [ASTROPHYS] A sudden increase in the radio-frequency energy radiated by the sun, generally associated with visible solar flares. { 'sō·lər 'bərst }

solar calendar [ASTRON] A calendar based on the time period known as the tropical year, which has 365.24220 days. { 'sō·lər 'kal·ən·dər }

solar cavity *See* heliosphere. { 'sō·lər 'kav·əd·ē }

solar corona [ASTRON] The upper, rarefied solar atmosphere which becomes visible around the darkened sun during a total solar eclipse. Also known as corona. { 'sō·lər kə'rō·nə }

solar cosmic rays *See* energetic solar particles. { 'sō·lər 'käz·mik 'rāz }

solar cycle [ASTRON] The periodic change in the number of sunspots; the cycle is taken as the interval between successive minima and is about 11.1 years. { 'sō·lər 'sī·kəl }

solar day [ASTRON] A time measurement, the duration of one rotation of the earth on its axis with respect to the sun; this may be a mean solar day or an apparent solar day as the reference is to the mean sun or apparent sun. { 'sō·lər 'dā }

solar eclipse [ASTRON] An eclipse that takes place when the new moon passes between the earth and the sun and the shadow formed reaches the earth; may be classified as total, partial, or annular. { 'sō·lər i'klips }

solar energy [ASTROPHYS] The energy transmitted from the sun in the form of electromagnetic radiation. { 'sō·lər 'en·ər·jē }

solar faculae [ASTRON] Bright streaks or regions on the surface of the sun, especially near solar sunspots. { 'sō·lər 'fak·yə,lē }

solar flare [ASTROPHYS] An abrupt increase in the intensity of the H-α and other emission near a sunspot region; the brightness may be many times that of the associated plage. { 'sō·lər 'fler }

solar flux unit [ASTRON] A unit of solar radio emission per unit frequency interval, equal to 10^{-22} watt per square meter per hertz at the earth. { 'sō·lər ¦fləks ,yü·nət }

solar heat exchanger drive [AERO ENG] A proposed method of spacecraft propulsion in which solar radiation is focused on an area occupied by a boiler to heat a working fluid that is expelled to produce thrust directly. Abbreviated SHED. { 'sō·lər 'hēt iks¦chān·jər ,drīv }

solar magnetic field [ASTROPHYS] The magnetic field that pervades the ionized and highly conducting gas composing the sun. { 'sō·lər mag'ned·ik 'fēld }

solar month [ASTRON] A time interval equal to one-twelfth of the solar year. { 'sō·lər 'mənth }

solar motion [ASTRON] The two main motions of the sun: relative motion with respect to the neighboring stars, or motion due to the rotation of the Milky Way of which the sun is a part. { 'sō·lər 'mō·shən }

solar nebula [ASTRON] The rotating flattened cloud of gas and dust from which the sun and the rest of the bodies in the solar system formed, about 4.56×10^9 years ago. { ¦sō·lər 'neb·yə·lə }

solar neutrino [ASTROPHYS] A neutrino produced in a nuclear reaction inside the sun. { 'sō·lər nü'trē·nō }

solar neutrino problem [ASTROPHYS] The difficulty of understanding why the observed flux of neutrinos from the sun is significantly lower than predicted by standard solar models; this discrepancy suggests the existence of some type of neutrino oscillation. { ¦sō·lər nü'trē·nō ‚präb·ləm }

solar neutrino unit [ASTROPHYS] A unit for measuring the capture rate of neutrinos emanating from the sun, equal to 10^{-36} per second per atom. Abbreviated SNU. { 'sō·lər nü'trē·nō ‚yü·nət }

solar nutation [ASTRON] Nutation caused by the change in declination of the sun. { 'sō·lər nü'tā·shən }

solar orbit [ASTRON] An orbit of a planet or other celestial body or satellite about the sun. { 'sō·lər 'òr·bət }

solar parallax [ASTRON] The sun's mean equatorial horizontal parallax p, which is the angle subtended by the equatorial radius r of the earth at mean distance a of the sun. { 'sō·lər 'par·ə‚laks }

solar phase angle [ASTRON] The angular distance between the earth and the sun at a specified planet. { 'sō·lər 'fāz ‚aŋ·gəl }

solar physics [ASTROPHYS] The scientific study of all physical phenomena connected with the sun; it overlaps with geophysics in the consideration of solar-terrestrial relationships, such as the connection between solar activity and auroras. { 'sō·lər 'fiz·iks }

solar probe [AERO ENG] A space probe whose trajectory passes near the sun so that instruments on board may detect and transmit back to earth data about the sun. { 'sō·lər 'prōb }

solar prominence [ASTRON] Sheets of luminous gas emanating from the sun's surface; they appear dark against the sun's disk but bright against the dark sky, and occur only in regions of horizontal magnetic fields. { 'sō·lər 'präm·ə·nəns }

solar propulsion [AERO ENG] Spacecraft propulsion with a system composed of a type of solar engine. { 'sō·lər prə'pəl·shən }

solar radiation [ASTROPHYS] The electromagnetic radiation and particles (electrons, protons, and rarer heavy atomic nuclei) emitted by the sun. { 'sō·lər ‚rād·ē'ā·shən }

solar radio emission [ASTROPHYS] Radio-frequency electromagnetic radiation emitted from the sun, and increasing greatly in intensity during sunspots and flares. { 'sō·lər 'rād·ē·ō i‚mish·ən }

solar rocket [AERO ENG] A rocket designed to carry instruments to measure and transmit parameters of the sun. { 'sō·lər 'räk·ət }

solar sail [AERO ENG] A surface of a highly polished material upon which solar light radiation exerts a pressure. Also known as photon sail. { 'sō·lər 'sāl }

solar satellite [AERO ENG] A space vehicle designed to enter into orbit about the sun. Also known as sun satellite. { 'sō·lər 'sad·əl‚īt }

solar sector [ASTRON] A region of the solar wind in which one magnetic polarity predominates. { 'sō·lər 'sek·tər }

solar spectrum [ASTROPHYS] The spectrum of the sun's electromagnetic radiation extending over the whole electromagnetic spectrum, from wavelengths of 10^{-9} centimeter to 30 kilometers. { 'sō·lər 'spek·trəm }

solar spicule See spicule. { 'sō·lər 'spik·yül }

solar system [ASTRON] The sun and the celestial bodies moving about it; the bodies are planets, satellites of the planets, asteroids, comets, and meteor swarms. { 'sō·lər ‚sis·təm }

solar time [ASTRON] Time based on the rotation of the earth relative to the sun. { 'sō·lər 'tīm }

solar tower [ASTRON] A tower which has a coelostat mounted on top to reflect the sun's light vertically downward so that it may be studied with a spectroheliograph or other astronomical instrument at the bottom of the tower. { 'sō·lər ˌtaü·ər }

solar turboelectric drive [AERO ENG] A proposed method of spacecraft propulsion in which solar radiation is focused on an area occupied by a boiler to heat a working fluid that drives a turbine generator system, producing electrical energy. Abbreviated STED. { 'sō·lər ˌtər·bō·i¦lek·trik 'drīv }

solar-type star [ASTRON] Any of the stars (yellow stars) of spectral type G, so called because the sun is in this class. { 'sō·lər ¦tīp 'stär }

solar ultraviolet radiation [ASTROPHYS] That portion of the sun's electromagnetic radiation that has wavelengths from about 400 to about 4 nanometers; this radiation may sufficiently ionize the earth's atmosphere so that propagation of radio waves is affected. { 'sō·lər ¦əl·trə¦vī·lət ˌrād·ē'ā·shən }

solar units [ASTROPHYS] A set of units for measuring properties of stars, in which properties of the sun such as mass, diameter, density, and luminosity are set equal to unity. { 'sō·lər ˌyü·nəts }

solar velocity [ASTRON] The sun's velocity with respect to the local standard of rest. { 'sō·lər və'läs·əd·ē }

solid-propellant rocket engine [AERO ENG] A rocket engine fueled with a solid propellant; such motors consist essentially of a combustion chamber containing the propellant, and a nozzle for the exhaust jet. { 'säl·əd prə¦pel·ənt 'räk·ət ˌen·jən }

solid rocket [AERO ENG] A rocket that is propelled by a solid-propellant rocket engine. { 'säl·əd 'räk·ət }

solstice [ASTRON] The two days (actually, instants) during the year when the earth is so located in its orbit that the inclination (about 23¹/₂°) of the polar axis is toward the sun; the days are June 21 for the North Pole and December 22 for the South Pole; because of leap years, the dates vary a little. { 'sälz·təs }

solstitial colure [ASTRON] That great circle of the celestial sphere through the celestial poles and the solstices. { sälz'tish·əl kə'lür }

solstitial points [ASTRON] Those points of the ecliptic that are 90° from the equinoxes north or south at which the greatest declination of the sun is reached. { sälz'tish·əl 'póins }

Sombrero galaxy [ASTRON] A spiral galaxy in the constellation Virgo that is seen nearly edge-on, having a recession velocity of approximately 910 kilometers per second (565 miles per second). { səm'brer·ō 'gal·ik·sē }

Sothic cycle [ASTRON] A time period of about 1460 years; this cycle is such that the New Year of the calendar used in ancient Egypt was in error by a whole year because the adopted year of 365 days is about a quarter of a day shorter than the mean solar year. { 'sä·thik ˌsī·kəl }

source function [ASTROPHYS] The emissivity of a stellar or other radiating material divided by its opacity. { 'sórs ˌfəŋk·shən }

Southern Cross See Crux. { 'səth·ərn 'krós }

Southern Crown See Corona Australis. { 'səth·ərn 'kraün }

Southern Fish See Piscis Austrinus. { 'səth·ərn 'fish }

Southern Triangle See Triangulum Australe. { 'səth·ərn 'trī,aŋ·gəl }

south point [ASTRON] That imaginary point on the celestial sphere at which the meridian intersects the horizon; it is due south of the observer. { 'saüth 'póint }

south polar distance [ASTRON] The angular distance between a celestial object and the south celestial pole. { 'saüth 'pō·lər 'dis·təns }

south tropical disturbance [ASTRON] An elongated dark band seen on the surface of Jupiter at about the latitude of the Great Red Spot; it has at times exceeded 180° of longitude in length and, like the Red Spot, appears and disappears intermittently. { 'saüth 'trap·ə·kəl di'stər·bəns }

space [ASTRON] **1.** Specifically, the part of the universe lying outside the limits of the earth's atmosphere. **2.** More generally, the volume in which all celestial bodies, including the earth, move. { spās }

space capsule [AERO ENG] A container, manned or unmanned, used for carrying out an experiment or operation in space. { 'spās ,kap·səl }

spacecraft [AERO ENG] Devices, crewed and uncrewed, which are designed to be placed into an orbit about the earth or into a trajectory to another celestial body. Also known as space ship; space vehicle. { 'spās,kraft }

spacecraft launching [AERO ENG] The setting into motion of a space vehicle with sufficient force to cause it to leave the earth's atmosphere. { 'spās,kraft ,lónch·iŋ }

spacecraft propulsion [AERO ENG] The use of rocket engines to accelerate space vehicles. { 'spās,kraft prə,pəl·shən }

space environment [ASTRON] The environment encountered by vehicles and living creatures in space, characterized by absence of atmosphere. { 'spās in,vi·ərn·mənt }

space flight [AERO ENG] Travel beyond the earth's sensible atmosphere; space flight may be an orbital flight about the earth or it may be a more extended flight beyond the earth into space. { 'spās ,flīt }

space-flight trajectory [AERO ENG] The track or path taken by a spacecraft. { 'spās ¦flīt trə'jek·trē }

space mission [AERO ENG] A journey by a vehicle, manned or unmanned, beyond the earth's atmosphere, usually for the purpose of collecting scientific data. { 'spās ,mish·ən }

space motion [ASTRON] Motion of a celestial body through space. { 'spās ,mō·shən }

spaceport [AERO ENG] An installation used to test and launch spacecraft. { 'spās,pórt }

space power system [AERO ENG] An on-board assemblage of equipment to generate and distribute electrical energy on satellites and spacecraft. { 'spās 'paú·ər ,sis·təm }

space probe [AERO ENG] An instrumented vehicle, the payload of a rocket-launching system designed specifically for flight missions to other planets or the moon and into deep space, as distinguished from earth-orbiting satellites. { 'spās ,prōb }

space reconnaissance [AERO ENG] Reconnaissance of the surface of a planet from a space ship or satellite. { 'spās ri,kän·ə·səns }

space reddening [ASTRON] Reddening of light from distant stars caused by selective absorption of blue light by interstellar dust clouds. { 'spās ,red·ən·iŋ }

space research [AERO ENG] Research involving studies of all aspects of environmental conditions beyond the atmosphere of the earth. { 'spās ri,sərch }

space satellite [AERO ENG] A vehicle, crewed or uncrewed, for orbiting the earth. { 'spās ,sad·əl,īt }

space shuttle [AERO ENG] A reusable orbital spacecraft, designed to travel from the earth to an orbital trajectory and then to return. { 'spās ,shəd·əl }

space simulator [AERO ENG] **1.** Any device which simulates one or more parameters of the space environment and which is used to test space systems or components. **2.** Specifically, a closed chamber capable of reproducing approximately the vacuum and normal environments of space. { 'spās ,sim·yə,lād·ər }

space station [AERO ENG] An autonomous, permanent facility in space for the conduct of scientific and technological research, earth-oriented applications, and astronomical observations. { 'spās ,stā·shən }

space technology [AERO ENG] The systematic application of engineering and scientific disciplines to the exploration and utilization of outer space. { 'spās tek,näl·ə·jē }

space velocity [ASTRON] A star's true velocity with reference to the sun. { 'spās və,läs·əd·ē }

space walk [AERO ENG] The movement of an astronaut outside the protected environment of a spacecraft during a space flight; the astronaut wears a spacesuit. { 'spās ,wók }

specific impulse [AERO ENG] A performance parameter of a rocket propellant, expressed in seconds, equal to the thrust in pounds divided by the weight flow rate in pounds per second. Also known as specific thrust. { spə'sif·ik 'im,pəls }

spectral classification [ASTRON] A classification of stars by characteristics revealed by study of their spectra; the six classes B, A, F, G, K, and M include 99% of all known stars. { 'spek·trəl ˌklas·ə·fə'kā·shən }

spectral luminosity classification See MK system. { 'spek·trəl ˌlü·mə'näs·əd·ē ˌklas·ə·fə ˌkā·shən }

spectral type [ASTRON] A label used to indicate the physical and chemical characteristics of a star as indicated by study of the star's spectra; for example, the stars in the spectral type known as class B are blue-white, and are referred to as helium stars because the dominant lines in their spectra are the lines in helium spectra. { 'spek·trəl ˌtīp }

spectroheliogram [ASTRON] A photograph of the sun obtained by means of a spectroheliograph. { ¦spek·trō'hē·lē·ə¦gram }

spectroscopic binary star [ASTRON] A binary star that may be distinguished from a single star only by noting the Doppler shift of the spectral lines of one or both stars as they revolve about their common center of mass. { ¦spek·trə¦skäp·ik 'bī¦ner·ē 'stär }

spectroscopic parallax [ASTRON] Parallax as determined from examination of a stellar spectrum; critical spectral lines indicate the star's absolute magnitude, from which the star's distance, or parallax, can be deduced. { ¦spek·trə¦skäp·ik 'par·əˌlaks }

spectrum of turbulence [ASTROPHYS] A relationship between the size of turbulent eddies in the sun's atmosphere and their average speed. { 'spek·trəm əv 'tər·byə·ləns }

spectrum variable [ASTRON] A main-sequence star of spectral class A whose spectrum displays anomalously strong lines of metals and rare earths whose intensity varies by about 0.1 magnitude over periods from 1 to 25 days. { 'spek·trəm 'ver·ē·ə·bəl }

spheres of Eudoxus [ASTRON] A theory of Eudoxus from about 400 B.C.; the planets, sun, and moon were on a series of concentric spheres rotating inside one another on different axes. { 'sfirz əv yü'däk·səs }

spheroidal galaxy See elliptical galaxy. { sfir'ȯid·əl 'gal·ik·sē }

Spica [ASTRON] A blue-white dwarf star of stellar magnitude 1.0, 160 light-years from the sun, spectral classification B1-V, in the constellation Virgo; the star α Virginis. { 'spī·kə }

spicule [ASTRON] One of an irregular distribution of jets shooting up from the sun's chromosphere. Also known as solar spicule. { 'spik·yül }

spike nozzle [AERO ENG] A nozzle in which gas is initially directed radially inward toward the nozzle axis, and expansion occurs only outside the nozzle when the gas is directly exposed to the external environment. { 'spīk ˌnäz·əl }

spin rocket [AERO ENG] A small rocket that imparts spin to a larger rocket vehicle or spacecraft. { 'spin ˌräk·ət }

spin stabilization [AERO ENG] Directional stability of a spacecraft obtained by the action of gyroscopic forces which result from spinning the body about its axis of symmetry. { 'spin ˌstā·bə·ləˌzā·shən }

spin-up [ASTRON] A sudden increase in the pulse frequency of a pulsar. { 'spin·əp }

spiral arms [ASTRON] The shape of sections of certain galaxies called spirals; these sections are two so-called arms composed of stars, dust, and gas extending from the center of the galaxy and coiled about it. { 'spī·rəl 'ärmz }

spiral galaxy [ASTRON] A type of galaxy classified on the basis of appearance of its photographic image; this type includes two main groups: normal spirals with circular symmetry of the nucleus and of the spiral arms, and barred spirals in which the dominant form is a luminous bar crossing the nucleus with spiral arms starting at the ends of the bar or tangent to a luminous rim on which the bar terminates. { 'spī·rəl 'gal·ik·sē }

spiral ring structure [ASTRON] A lunar crater in which ridges spiral inward from the main wall across the floor. { 'spī·rəl 'riŋ ˌstrək·chər }

splashdown [AERO ENG] **1.** The landing of a spacecraft or missile on water. **2.** The moment of impact of a spacecraft on water. { 'splashˌdaůn }

sporadic meteor [ASTRON] A meteor which is not associated with one of the regularly recurring meteor showers or streams. { spə'rad·ik 'mēd·ē·ər }

Spörer minimum [ASTRON] A period of low sunspot activity that occurred between 1420 and 1570. { 'spər·ər 'min·ə·məm }

Spörer's law [ASTRON] A relationship to indicate the frequency of occurrence of sunspots and their progressive movement to lower latitudes on the sun. { 'spər·ərz ,lò }

spot group [ASTROPHYS] A complex formation of the sun's surface consisting of a sunspot with several umbrae surrounded by a single penumbra, or of several sunspots which are close together and clearly associated. { 'spät ,grüp }

spray [ASTROPHYS] An explosive release of gas in all directions from the sun's chromosphere, with velocities as high as 930 miles (1500 kilometers) per second, which normally occurs in the first minutes of a flare. { 'sprā }

spring [ASTRON] The period extending from the vernal equinox to the summer solstice; comprises the transition period from winter to summer. { spriŋ }

spring equinox See vernal equinox. { 'spriŋ 'ē·kwə,näks }

SS Cygni stars See U Geminorum stars. { ¦es¦es 'sig·nē ,stärz }

SS 433 [ASTRON] A stellar object that shows evidence of ejection of two narrow streams of cool gas traveling in opposite direction from a cool object at a velocity of almost one-quarter the speed of light; the beams execute a repeating, rotating pattern about the central object once every 164 days. { ,es,es ,fȯr,thərd·ē'thrē }

S star [ASTRON] A spectral classification of stars, comprising red stars with surface temperature of about 2200 K; prominent in the spectra is zirconium oxide. { 'es ,stär }

stage [AERO ENG] A self-propelled separable element of a rocket vehicle or spacecraft. { stāj }

staging [AERO ENG] The process or operation during the flight of a rocket vehicle whereby a full stage or half stage is disengaged from the remaining body and made free to decelerate or be propelled along its own flightpath. { 'stāj·iŋ }

standard coordinates [ASTRON] A coordinate system used to locate stars on a photographic plate, in which the coordinates are the differences in right ascension and declination between the position of each star and the assumed position of the plate center. { 'stan·dərd kō'ȯrd·ən·əts }

standard noon [ASTRON] Twelve o'clock standard time, or the instant the mean sun is over the upper branch of the standard meridian. { 'stan·dərd 'nün }

standard star [ASTRON] A star whose position or other data are precisely known so that it is used as a reference to calculate positions of other celestial bodies, or of objects on earth. { 'stan·dərd 'stär }

standard time [ASTRON] The mean solar time, based on the transit of the sun over a specified meridian, called the time meridian, and adopted for use over an area that is called a time zone. { 'stan·dərd 'tīm }

standstill [ASTRON] An interval in the cycle of a variable star during which its brightness remains nearly constant. { 'stan,stil }

star [ASTRON] A celestial body consisting of a large, self-luminous mass of hot gas held together by its own gravity; the sun is a typical star. { stär }

star atlas [ASTRON] A series of star maps for different times, for example, for each month; the maps are generally drawn to a small scale and in book form. { 'stär ,at·ləs }

starburst galaxy [ASTRON] A galaxy that is presently undergoing a period of intense star formation. { 'stär,bərst ,gal·ik·sē }

star catalog [ASTRON] A comprehensive tabulation of data concerning the stars listed; the data may include, for example, apparent positions, brightness, motions, parallaxes, and other properties of stars. { 'stär ¦kad·əl,äg }

star chart See star map. { 'stär ,chärt }

star cloud [ASTRON] An aggregation of thousands or of millions of stars spread over hundreds or thousands of light-years. { 'stär ,klaůd }

star cluster [ASTRON] A group of stars held together by gravitational attraction; the two chief types are open clusters (composed of from 12 to hundreds of stars) and globular clusters (composed of thousands to hundreds of thousands of stars). Also known as cluster. { 'stär ,kläs·tər }

star color [ASTRON] The color of a star as a function of its radiation and related to its temperature; colors range from blue-white to deep red. { 'stär ,kəl·ər }

star count [ASTRON] A count of stars on a photographic plate. { 'stär ,kaunt }

star day [ASTRON] The time period between two successive passages of a star across the meridian. { 'stär 'dā }

star density [ASTRON] The average number of stars in a unit volume of space. { 'stär ,den·səd·ē }

star drift [ASTRON] A description of two star groups in the Milky Way traveling through each other in opposite directions; individual stars have movements that are relative to each other. Also known as star stream. { 'stär ,drift }

star finder [ASTRON] A device such as a star map or celestial globe to facilitate the identification of stars. Also known as a star identifier. { 'stär ,fin·dər }

star globe *See* celestial globe. { 'stär ,glōb }

star group [ASTRON] A number of stars that move in the same general direction at the same time. { 'stär ,grüp }

star identifier *See* star finder. { 'stär ī,den·tə,fī·ər }

star map [ASTRON] A map indicating the relative apparent positions of the stars. Also known as star chart. { 'stär ,map }

star model *See* stellar model. { 'stär ,mäd·əl }

star motions [ASTRON] For the Milky Way, this includes rotation within the galaxy, motion which is described with respect to an external frame of reference; superposed on this systematic rotation are the individual motions of a star; each star moves in a somewhat elliptical orbit, with respect to the local standard of rest, the standard moving in a circular orbit around the galactic center. { 'stär ,mō·shənz }

star names [ASTRON] Nomenclature for the identification of stars; hundreds of stars have proper names that are traditional, for example, Betelgeuse; this star may be also identified as α Orionis (Alpha Orionis), α for its being the brightest visual star in the constellation Orion. { 'stär ,nāmz }

star place [ASTRON] The position of a star on the celestial sphere, usually measured by its right ascension and declination. { 'stär ,plās }

starspot [ASTRON] A region of reduced brightness of the surface of a star comparable to a sunspot on the Sun's surface. { 'stär,spät }

star stream *See* star drift. { 'stär ,strēm }

star streaming [ASTRON] A phenomenon that results from the mean random speeds of stars being different in different directions. { 'stär ,strēm·iŋ }

static firing [AERO ENG] The firing of a rocket engine in a hold-down position to measure thrust and to accomplish other tests. { 'stad·ik 'fīr·iŋ }

static test [AERO ENG] In particular, a test of a rocket or other device in a stationary or hold-down position, either to verify structural design criteria, structural integrity, and the effects of limit loads, or to measure the thrust of a rocket engine. { 'stad·ik 'test }

static universe [ASTRON] A postulated universe that has a finite static volume and is closed. { 'stad·ik 'yü·nə,vərs }

stationary orbit [AERO ENG] A circular, equatorial orbit in which the satellite revolves about the primary body at the angular rate at which the primary body rotates on its axis; from the primary body, the satellite thus appears to be stationary over a point on the primary body; a stationary orbit must be synchronous, but the reverse need not be true. { 'stā·shə,ner·ē 'òr·bət }

stationary point [ASTRON] A point at which a planet's apparent motion changes from direct to retrograde motion, or vice versa. { 'stā·shə,ner·ē 'pòint }

stationary satellite [AERO ENG] A satellite in a stationary orbit. { 'stā·shə,ner·ē 'sad·əl,īt }

stationkeeping [AERO ENG] Keeping a satellite in geosynchronous orbit within assigned boundaries, typically within a few tenths of a degree of longitude, generally with the assistance of jet thrusters. { 'stā·shən,keep·iŋ }

statistical parallax [ASTRON] The mean parallax of a collection of stars that are all at approximately the same distance, as determined from their radial velocities and proper motions. { stə'tis·tə·kəl 'par·ə͵laks }

stay time [AERO ENG] In rocket engine usage, the average value of the time spent by each gas molecule or atom within the chamber volume. { 'stā ͵tīm }

steady-state theory [ASTRON] A cosmological theory which holds that the average density of matter does not vary with space or time in spite of the expansion of the universe; this requires that matter be continuously created. { 'sted·ē ͵stāt 'thē·ə·rē }

stellar [ASTRON] Relating to or consisting of stars. { 'stel·ər }

stellar association [ASTRON] A loose grouping of stars which may have had a common origin. { 'stel·ər ə͵sō·sē'ā·shən }

stellar atmosphere [ASTRON] The envelope of gas and plasma surrounding a star; consists of about 90% hydrogen atoms and 9% helium atoms, by number of atoms. { 'stel·ər 'at·mə͵sfir }

stellar evolution [ASTROPHYS] The changes in spectrum and luminosity that take place in the life of a star. { 'stel·ər ͵ev·ə'lü·shən }

stellar flare [ASTRON] Ejection of material from a star in an eruption that may last from a few minutes to an hour or more. { 'stel·ər 'fler }

stellar light [ASTRON] The part of the background illumination of the night sky that results from direct light from stars too faint to be visible to the unaided eye. { 'stel·ər 'līt }

stellar luminosity [ASTRON] A star's brightness; it is measured either in ergs per second or in units of solar luminosity or in absolute magnitude. { 'stel·ər lü·mə'näs·əd·ē }

stellar magnetic field [ASTROPHYS] A magnetic field, generally stronger than the earth's magnetic field, possessed by many stars. { 'stel·ər mag'ned·ik 'fēld }

stellar magnitude *See* magnitude. { 'stel·ər 'mag·nə͵tüd }

stellar mass [ASTROPHYS] The mass of a star, usually expressed in terms of the sun's mass. { 'stel·ər 'mas }

stellar model [ASTROPHYS] A mathematical characterization of the internal properties of a star. Also known as star model. { 'stel·ər 'mäd·əl }

stellar parallax [ASTRON] The subtended angle at a star formed by the mean radius of the earth's orbit; it indicates distance to the star. { 'stel·ər 'par·ə͵laks }

stellar photometry [ASTRON] The measurement of the brightness of stars. { 'stel·ər phə'täm·ə·trē }

stellar population [ASTRON] Either of two classes of stars, termed population I and population II; population I are relatively young stars, found in the arms of spiral galaxies, especially the blue stars of high luminosity; population II stars are the much older, more evolved stars of lower metallic content; many high luminosity red giants and many variable stars are members of population II. { 'stel·ər ͵päp·yə'lā·shən }

stellar pulsation [ASTROPHYS] Expansion of a star followed by contraction so that its surface temperature and intrinsic brightness undergo periodic variation. { 'stel·ər pəl'sā·shən }

stellar rotation [ASTRON] Axial rotation of stars; surface rotational equatorial velocities of stars range from a few to 500 kilometers per second. { 'stel·ər rō'tā·shən }

stellar scintillation *See* astronomical scintillation. { 'stel·ər ͵sint·əl'ā·shən }

stellar spectroscopy [ASTRON] The techniques of obtaining spectra of stars and their study. { 'stel·ər spek'träs·kə·pē }

stellar spectrum [ASTRON] The spectrum of a star normally obtained with a slit spectrograph by black-and-white photography; the spectrum of a star in a large majority of cases shows absorption lines superposed on a continuous background. { 'stel·ər 'spek·trəm }

stellar structure [ASTROPHYS] The mathematical study of a rotating, chemically homogeneous mass of gas held together by its own gravitation; a representative model of the observable properties of a star; thermonuclear reactions are postulated to be the main source of stellar energy. { 'stel·ər ͵strək·chər }

stellar system [ASTRON] A gravitational system of stars. { 'stel·ər ͵sis·təm }

stellar temperature |ASTROPHYS| Any temperature above several million degrees, such as occurs naturally in the interior of the sun and other stars.   { 'stel·ər 'tem·prə·chər }

stellar wind |ASTRON| The flow of ionized gas from the surface of a star into interstellar space.   { 'stel·ər 'wind }

Stephano |ASTRON| A small satellite of Uranus in a retrograde orbit with a mean distance of 4,940,000 miles (7,950,000 kilometers), eccentricity of 0.053, and sidereal period of 1.84 years.   { 'stef·ə·nō or stə'fä·nō }

Stephan's Quintet |ASTRON| A group of five galaxies which lie close together, one of which has widely divergent red shifts.   { 'stef·ənz kwin'tet }

Strömgren four-color index See uvby system.   { 'strəm·grən ¦fȯr ¦kəl·ər 'in,deks }

Strömgren radius |ASTRON| The radius of a sphere surrounding the central star of an emission nebula within which the hydrogen is nearly completely ionized.   { 'strəm· grən ,rād·ē·əs }

Strömgren sphere |ASTRON| An approximately spherical region of ionized hydrogen that surrounds a hot star.   { 'strəm·grən ,sfir }

structure |AERO ENG| The construction or makeup of an airplane, spacecraft, or missile, including that of the fuselage, wings, empennage, nacelles, and landing gear, but not that of the power plant, furnishings, or equipment.   { 'strək·chər }

S-type asteroid |ASTRON| A type of asteroid whose surface is reddish and of moderate albedo, containing pyroxene and olivine silicates, probably mixed with metallic iron, similar to stony iron meteorites.   { 'es ¦tīp 'as·tə,rȯid }

S-type symbiotic star |ASTRON| A member of a class of symbiotic stars that emit infrared radiation typical of red-giant atmospheres; they are relatively dust-free and have binary periods shorter than 20 years.   { ¦es,tīp ,sim·bē,äd·ik 'stär }

subastral point See substellar point.   { ¦səb'as·trəl 'pȯint }

subcluster |ASTRON| One of the several distinct clumps of galaxies that often compose an irregular cluster.   { 'səb,kləs·tər }

subdwarf star |ASTRON| An intermediate star type; luminosity is between that of main sequence stars and the white dwarf stars on the Hertzsprung-Russell diagram; spectral classes F, G, and K are most numerous.   { 'səb,dwȯrf 'stär }

subdwarf symbiotic |ASTRON| A type of symbiotic star consisting of a combination of a cool red giant with a small hot subdwarf star, the latter probably the inner core of a former giant or super giant which has shed its outer envelope and is now contracting to become a white dwarf.   Also known as planetary nebula symbiotic.   { 'səb,dwȯrf ,sim·bē'äd·ik }

subgiant CH star |ASTRON| A type of star that resembles the CH stars and barium stars but is less luminous and somewhat hotter, with some members of the class lying on or near the main sequence.   { ¦səb¦jī·ənt ¦sē'āch ,stär }

subgiant star |ASTRON| A member of the family of stars whose luminosity is intermediate between giants and the main sequence in the Hertzsprung-Russell diagram; spectral classes G and K are most frequent.   { ¦səb'jī·ənt 'stär }

subluminous star |ASTRON| A star that is fainter than those of the same color on the main sequence.   { səb'lüm·ə·nəs 'stär }

sublunar point |ASTRON| The moon's geographic zenith position at any particular moment in time.   { ¦səb'lü·nər 'pȯint }

submillimeter astronomy |ASTRON| Astronomical observations carried out in the region of the electromagnetic spectrum with wavelengths from approximately 0.3 to 1.0 millimeter.   { ,səb¦mil·ə,mēd·ər ə'strän·ə·mē }

subpulse |ASTRON| The weaker component of a pulsar's periodic emission.   { 'səb,pəls }

subsatellite |AERO ENG| An object that is carried into orbit by, and subsequently released from, an artificial satellite.   { 'səb,sad·əl,īt }

subsolar point |ASTRON| The sun's zenith geographic position at any particular moment in time.   { ¦səb'sō·lər 'pȯint }

substellar object See brown dwarf.   { ,səb,stel·ər 'äb,jekt }

substellar point [ASTRON] The geographical position of a star; that point on the earth at which the star is in the zenith at a specified time. Also known as subastral point. { ¦səb'stel·ər 'pȯint }

summer [ASTRON] The period from the summer solstice to the autumnal equinox; popularly and for most meteorological purposes, it is taken to include June through August in the Northern Hemisphere, and December through February in the Southern Hemisphere. { 'səm·ər }

summer noon See daylight saving noon. { 'səm·ər 'nün }

summer solstice [ASTRON] **1.** The sun's position on the ecliptic when it reaches its greatest northern declination. Also known as first point of Cancer. **2.** The date, about June 21, on which the sun has its greatest northern declination. { 'səm·ər 'säl·stəs }

summer time See daylight saving time. { 'səm·ər 'tīm }

sun [ASTRON] The star about which the earth revolves; it is a globe of gas 8.65 × 10^5 miles (1.392 × 10^6 kilometers) in diameter, held together by its own gravity; thermonuclear reactions take place in the deep interior of the sun converting hydrogen into helium releasing energy which streams out. Also known as Sol. { sən }

sun dog See parhelion. { 'sən ˌdȯg }

sun-grazing comet [ASTRON] A comet whose orbit causes it to either collide with the sun or completely disintegrate in the outer solar atmosphere. { 'sən ¦grāz·iŋ 'käm·ət }

sunrise [ASTRON] The exact moment the upper limb of the sun appears above the horizon. { 'sən,rīz }

sunset [ASTRON] The exact moment the upper limb of the sun disappears below the horizon. { 'sən,set }

sunshine [ASTRON] Direct radiation from the sun, as opposed to the shading of a location by clouds or by other obstructions. { 'sən,shīn }

sunspot [ASTRON] A dark area in the photosphere of the sun caused by a lowered surface temperature. { 'sən,spät }

sunspot cycle [ASTRON] Variation of the size and number of sunspots in an 11-year cycle which is shared by all other forms of solar activity. { 'sən,spät ˌsī·kəl }

sunspot maximum [ASTRON] The time in the solar cycle when the number of sunspots reaches a maximum value. { 'sən,spät 'mak·sə·məm }

sunspot number See relative sunspot number. { 'sən,spät ˌnəm·bər }

sunspot relative number See relative sunspot number. { 'sən,spät 'rel·əd·iv ˌnəm·bər }

sun's way [ASTRON] The path of the solar system through space. { 'sənz 'wā }

sun-synchronous orbit [AERO ENG] An earth orbit of a spacecraft so that the craft is always in the same direction relative to that of the sun; as a result, the spacecraft passes over the equator at the same spots at the same times. { 'sən ¦siŋ·krə·nəs 'ȯr·bət }

Sunyaev-Zeldovich effect [ASTRON] A change in the spectrum of the cosmic background radiation in the direction of galaxy clusters, whereby the radiation is less intense at wavelengths longer than 1.4 millimeters and more intense at shorter wavelengths as the result of hot gas in the clusters. { sún¦yä,ef zel'dō,vich i,fekt }

supercluster [ASTRON] An association of galaxy clusters and groups, typically composed of a few rich clusters and many poorer groups and isolated galaxies. { 'sü·pər,kləs·tər }

superdense state [ASTROPHYS] An extremely compact state of matter in which protons and electrons are pressed together to form neutrons, as in a neutron star. { ¦sü·pər¦dens 'stāt }

superdense theory See big bang theory. { ¦sü·pər'dens ˌthē·ə·rē }

supergalaxy [ASTRON] A hypothetical very large group of galaxies which together fill an ellipsoidal space. { ¦sü·pər'gal·ik·sē }

supergiant star [ASTRON] A member of the family containing the intrinsically brightest stars, populating the top of the Hertzsprung-Russell diagram; supergiant stars occur at all temperatures from 30,000 to 3000 K and have luminosities from 10^4 to 10^6 times that of the sun; the star Betelgeuse is an example. { ¦sü·pər'gī·ənt 'stär }

supergranulation [ASTRON] A system of convective cells, with typical diameters of 12,000 miles (20,000 kilometers), that cover the sun's surface. { ˌsü·pərˌgran·yə'lā·shən }

supergranulation cells [ASTRON] Convective cells in the solar photosphere with primarily horizontal flow and diameters of about 20,000 miles (30,000 kilometers). Also known as supergranules. { ˈsü·pərˌgran·yə'lā·shən ˌselz }

supergranules See supergranulation cells. { ˈsü·pər'gran·yülz }

superior conjunction [ASTRON] A conjunction when an astronomical body is opposite the earth on the other side of the sun. { sə'pir·ē·ər kən'jəŋk·shən }

superior planet [ASTRON] Any of the planets that are farther than the earth from the sun; includes Mars, Jupiter, Saturn, Uranus, Neptune, and Pluto. { sə'pir·ē·ər 'plan·ət }

superior transit See upper transit. { sə'pir·ē·ər 'tran·zət }

super-Jupiter See brown dwarf. { ˈsü·pər 'jü·pəd·ər }

superluminal motion [ASTRON] Apparent proper motion exceeding the velocity of light in an astronomical object. { ˌsü·pərˌlim·ə·nəl 'mō·shən }

superluminal radio source [ASTRON] A radio source whose velocity appears to exceed that of light. { ˈsü·pərˌlüm·ən·əl 'rād·ē·ō ˌsórs }

supermassive star [ASTRON] A star with a mass exceeding about 50 times that of the sun. { ˈsü·pər'mas·iv 'stär }

super-metal-rich star [ASTRON] **1.** A low-luminosity giant star of spectral class K, strongly enhanced cyanogen radical (CN) bands, and apparently strong metal lines. **2.** A star that is significantly richer in metals than those of the Hyades. { 'sü·pər ˈmed·əl ˌrich 'stär }

supernova [ASTRON] A star that suddenly bursts into very great brilliance as a result of its blowing up; it is orders of magnitude brighter than a nova. { ˈsü·pər'nō·və }

supernova remnant [ASTRON] A nebula consisting of an expanding shell of gas that has been ejected by a supernova. Abbreviated SNR. { ˈsü·pər'nō·və 'rem·nənt }

superthermal particles [ASTRON] Particles in the solar corona that have been accelerated by magnetic energy dissipation to very high energies, from 10^2 to 10^6 times the mean thermal energy of particles in the coronal gas. { ˈsü·pərˌthər·məl 'pärd·ə·kəlz }

superwind [ASTRON] An energetic outflow of hot material along the minor axis of some starburst galaxies, detected in x-rays and the emission lines of hydrogen. { 'sü·pərˌwind }

surge [ASTROPHYS] An unusually violent solar prominence that usually accompanies a smaller flare, consisting of a brilliant jet of gas which shoots out into the solar corona with a speed on the order of 180 miles (300 kilometers) per second and reaches a height on the order of 60,000 miles (100,000 kilometers). { sərj }

sustainer rocket engine [AERO ENG] A rocket engine that maintains the velocity of a rocket vehicle once it has achieved its programmed velocity by use of a booster or other engine. { sə'stān·ər 'räk·ət ˌen·jən }

Swan See Cygnus. { swän }

Swann bands [ASTROPHYS] Particular bands seen in the visible spectra of comets; they arise from the presence of dimeric carbon (C_2) in the comet's tail. { swän ˌbanz }

Swan Nebula See Omega Nebula. { 'swän 'neb·yə·lə }

swing-around trajectory [AERO ENG] A planetary round-trip trajectory which requires minimal propulsion at the destination planet, but instead uses the planet's gravitational field to effect the bulk of the necessary orbit change to return to earth. { 'swiŋ əˌraúnd trə'jek·trē }

Swordfish See Dorado. { 'sórdˌfish }

Sycorax [ASTRON] A small satellite of Uranus in a retrograde orbit with a mean distance of 7,550,000 miles (12,150,000 kilometers), eccentricity of 0.504, and sidereal period of 3.53 years. { sī 'kórˌaks }

symbiotic nova See novalike symbiotic. { ˌsim·bē'äd·ik 'nō·və }

symbiotic objects [ASTRON] Stars whose spectra have characteristics of two disparate spectral classes. { ˌsim·bē'äd·ik 'äbˌjeks }

symbiotic star [ASTRON] A stellar object whose optical spectrum displays features indicative of two very different stellar regimes: a stellar spectrum whose flux distribution and absorption lines suggest the presence of a cool star, and emission lines which can be formed only in a much hotter medium. { ,sim·bē'äd·ik 'stär }

synchrone [ASTRON] The geometrical locus of the dust grains ejected from the nucleus of a comet at the same time and having any value of beta. { 'siŋ,krōn }

synchronous orbit [AERO ENG] **1.** An orbit in which a satellite makes a limited number of equatorial crossing points which are then repeated in synchronism with some defined reference (usually earth or sun). **2.** Commonly, the equatorial, circular, 24-hour case in which the satellite appears to hover over a specific point of the earth. { 'siŋ·krə·nəs 'ȯr·bət }

synchronous rotation [ASTRON] The rotation of a planet or satellite whose period is equal to its orbital period. { 'siŋ·krə·nəs rō'tā·shən }

syndyne [ASTRON] The geometrical locus of the dust grains ejected from the nucleus of a comet continuously and having a particular value of beta. { 'sin,dīn }

synergic curve [AERO ENG] A curve plotted for the ascent of a rocket, space-air vehicle, or space vehicle, calculated to give optimum fuel economy and optimum velocity. { sə'nər·jik 'kərv }

synodic [ASTRON] Referring to conjunction of celestial bodies. { sə'näd·ik }

synodic month [ASTRON] A month based on the moon's phases. { sə'näd·ik 'mənth }

synodic period [ASTRON] The time period between two successive astronomical conjunctions of the same celestial objects. { sə'näd·ik 'pir·ē·əd }

syzygy [ASTRON] **1.** One of the two points in a celestial object's orbit where it is in conjunction with or opposition to the sun. **2.** Those points in the moon's orbit where the moon, earth, and sun are in a straight line. **3.** The alignment of any three objects within the solar system, or within any other system of objects in orbit about a star. { 'siz·ə·jē }

T

tachocline [ASTRON] A zone of radial shear in the interior of the sun that is located at the base of the convection zone at a distance from the sun's center equal to about 0.7 of the sun's radius, and represents an abrupt transition from the rotation rates in the convection zone to those in the underlying layers. { 'tak·ə,klīn }

tail [AERO ENG] **1.** The rear part of a body, as of an aircraft or a rocket. **2.** The tail surfaces of an aircraft or a rocket. [ASTRON] The part of a comet that extends from the coma in a direction opposite to the sun; it consists of dust and gas that have been blown away from the coma by the solar wind and the sun's radiation pressure. { tāl }

tail fin [AERO ENG] A fin at the rear of a rocket or other body. { 'tāl ,fin }

tail surface [AERO ENG] A stabilizing or control surface in the tail of an aircraft or missile. { 'tāl ,sər·fəs }

takeoff [AERO ENG] Ascent of an aircraft or rocket at any angle, as the action of a rocket vehicle departing from its launch pad or the action of an aircraft as it becomes airborne. { 'tāk,óf }

takeoff weight [AERO ENG] The weight of an aircraft or rocket vehicle ready for takeoff, including the weight of the vehicle, the fuel, and the payload. { 'tāk,óf ,wāt }

tandem [AERO ENG] The fore and aft configuration used in boosted missiles, long-range ballistic missiles, and satellite vehicles; stages are stacked together in series and are discarded at burnout of the propellant for each stage. { 'tan·dəm }

Tarantula *See* Loop Nebula. { tə'ran·chə·lə }

T association [ASTRON] An association that includes many T Tauri stars. { 'tē ə,sō·sē,ā·shən }

Taurid meteor [ASTRON] A meteor shower occurring from about October 26 to November 16 in the Northern Hemisphere, with the maximum occurring about the first week in November; the radiant lies in the constellation Taurus. { 'tór·əd 'mēd·ē·ər }

Taurus [ASTRON] A northern constellation; right ascension 4 hours, declination 15° north; it includes the star Aldebaran, useful in navigation. Also known as Bull. { 'tór·əs }

Taurus A [ASTRON] A strong, discrete radio source in the constellation Taurus, associated with the Crab Nebula. { 'tór·əs 'ā }

Taurus cluster [ASTRON] A cluster of stars observed in the region of the constellation Taurus; it is about 130 light-years (1.23×10^{18} meters) from the sun, and 58 light-years (5.49×10^{17} meters) in diameter. { 'tór·əs 'kləs·tər }

Taurus dark cloud [ASTRON] A large, relative nearby aggregate of dust and gas in which star formation is taking place. { 'tór·əs 'därk 'klaúd }

Telescope *See* Telescopium. { 'tel·ə,skōp }

telescopic comet [ASTRON] A comet in which only the coma is observed. { ¦tel·ə¦skäp·ik 'käm·ət }

Telescopium [ASTRON] A constellation, right ascension 19 hours, declination 50° south. Also known as Telescope. { ,tel·ə'skō·pē·əm }

Telesto [ASTRON] A small, irregularly shaped satellite of Saturn that librates about the trailing Lagrangian point of Tethys's orbit. { te'les·tō }

terminator [ASTRON] The line of demarcation between the dark and light portions of the moon or planets. { 'tər·mə,nād·ər }

terra [ASTRON] A bright upland or mountainous region on the surface of the moon, characterized by a lighter color than that of a mare, a relatively high albedo, and a rough texture formed by large intersecting or overlapping craters. { 'ter·ə }

terrestrial planet [ASTRON] One of the four small planets near the sun (Earth, Mercury, Venus, and Mars). { tə'res·trē·əl 'plan·ət }

test firing [AERO ENG] The firing of a rocket engine, either live or static, with the purpose of making controlled observations of the engine or of an engine component. { 'test ‚fīr·iŋ }

test flight [AERO ENG] A flight to make controlled observations of the operation or performance of an aircraft or a rocket, or of a component of an aircraft or rocket. { 'test ‚flīt }

Tethys [ASTRON] A satellite of the planet Saturn having a diameter of about 660 miles (1060 kilometers). { 'tē·thəs }

Thalassa [ASTRON] A satellite of Neptune orbiting at a mean distance of 31,000 miles (50,000 kilometers) with a period of 7.5 hours, and with a diameter of about 50 miles (80 kilometers). { thə'las·ə }

Tharsis rise [ASTRON] An uplifted portion of the surface of Mars that stands several kilometers above the mean elevation of the planet and affects approximately one-quarter of the entire surface. { 'thär·səs ‚rīz }

Thebe [ASTRON] A small satellite of Jupiter, having an orbital radius of 137,900 miles (221,900 kilometers), and a radius of 28–34 miles (48–55 kilometers). Also known as Jupiter XIV. { 'thē·bē }

Themis [ASTRON] An asteroid with a diameter of roughly 129 miles (207 kilometers), mean distance from the sun of 3.13 astronomical units, and C-type surface composition. { 'thē·məs }

thermal barrier [AERO ENG] A limit to the speed of airplanes and rockets in the atmosphere imposed by heat from friction between the aircraft and the air, which weakens and eventually melts the surface of the aircraft. Also known as heat barrier. { 'thər·məl 'bar·ē·ər }

thermal phase See gradual phase. { 'thər·məl ‚fāz }

thermal time scale See Kelvin time scale. { 'thər·məl 'tīm ‚skāl }

thick disk [ASTRON] A component of the galactic stellar population that extends over distances up to 1500 parsecs from the galactic plane, and is approximately 9–10 × 10^9 years old, intermediate in age between the thin disk and the halo. { ¦thik 'disk }

thin disk [ASTRON] The youngest component of the galactic stellar population, still actively forming massive stars from molecular clouds and confined to within about 350 parsecs (1100 light-years) of the galactic plane. { ¦thin 'disk }

30 Doradus See Loop Nebula. { ¦thər·dē də'rä·dəs }

Thisbe [ASTRON] An asteroid with a diameter of about 124 miles (200 kilometers), mean distance from the sun of 2.77 astronomical units, and C-type surface composition. { 'thiz·bē }

Thorne-Żytkow object [ASTRON] A hypothetical object that results when a star in its helium-burning, or red-giant, phase swallows a companion neutron star that had resulted from the evolution of a very massive star. { ¦thȯrn 'zhit‚kȯv ‚äb‚jekt }

three-alpha process [ASTROPHYS] A nuclear reaction in which three helium-4 nuclei (alpha particles) combine to form a carbon-12 nucleus, with the emission of a gamma ray; it converts helium into carbon in red giants. Also known as Salpeter process; triple-alpha process. { 'thrē ¦al·fə 'prä‚səs }

three-axis stabilization [AERO ENG] Directional stability of a spacecraft obtained without spin, usually with internal gyroscopes to maintain stability about each of three perpendicular axes. { ¦thrē ‚ak·səs ‚stā·bə·lə'zā·shən }

three-kiloparsec arm [ASTRON] A region approximately 3 kiloparsecs from the galactic center that displays strong absorption in the 21-centimeter line of atomic hydrogen. { 'thrē ¦kil·ō¦pär‚sek 'ärm }

throttling [AERO ENG] The varying of the thrust of a rocket engine during powered flight. { 'thräd·əl·iŋ }

thruster [AERO ENG] A control jet employed in spacecraft; an example would be one utilizing hydrogen peroxide. { 'thrəs·tər }

thrust horsepower [AERO ENG] **1.** The force-velocity equivalent of the thrust developed by a jet or rocket engine. **2.** The thrust of an engine-propeller combination expressed in horsepower; it differs from the shaft horsepower of the engine by the amount the propeller efficiency varies from 100%. { 'thrəst 'hȯrs,pau̇·ər }

thrust output [AERO ENG] The net thrust delivered by a jet engine, rocket engine, or rocket motor. { 'thrəst 'au̇t,pu̇t }

thrust-pound [AERO ENG] A unit of measurement for the thrust produced by a jet engine or rocket. { 'thrəst 'pau̇nd }

thrust power [AERO ENG] The power usefully expended on thrust, equal to the thrust (or net thrust) times airspeed. { 'thrəst ,pau̇·ər }

thrust terminator [AERO ENG] A device for ending the thrust in a rocket engine, either through propellant cutoff (in the case of a liquid) or through diverting the flow of gases from the nozzle. { 'thrəst ,tər·mə,nād·ər }

thrust-weight ratio [AERO ENG] A quantity used to evaluate engine performance, obtained by dividing the thrust output by the engine weight less fuel. { 'thrəst 'wāt ,rā·shō }

Thuban [ASTRON] A fourth-magnitude star of spectral class AO in the constellation Draco that was near the north celestial pole around 3000 B.C.; the star Alpha Draconis. { 'thü,ban }

Thule group [ASTRON] An accumulation of asteroids whose sidereal period of revolution is in the ratio 3/4 with that of Jupiter. { 'tü·lē ,grüp }

tidal radius [ASTRON] The distance from the center of a planet in formation at which the planet's gravitational attraction for nearby gas equals that of the Sun. { 'tīd·əl 'rād·ē·əs }

time diagram [ASTRON] A diagram in which the celestial equator appears as a circle, and celestial meridians and hour circles as radial lines; used to facilitate solution of time problems and other problems involving arcs of the celestial equator or angles at the pole, by indicating relations between various quantities involved; conventionally, the relationships are given as viewed from a point over the South Pole, in a westward direction or counterclockwise. Also known as diagram on the plane of the celestial equator; diagram on the plane of the equinoctial. { 'tīm ,dī·ə,gram }

time meridian [ASTRON] Any meridian used as a reference for reckoning time, particularly a zone or standard meridian. { 'tīm mə,rid·ē·ən }

time zone [ASTRON] To avoid the inconvenience of the continuous change of mean solar time with longitude, the earth is divided into 24 time zones, each about 15° wide and centered on standard longitudes, 0°, 15°, 30°, and so on; within each zone the time kept is the mean solar time of the standard meridian. { 'tīm ,zōn }

Titan [ASTRON] The largest satellite of Saturn, with a diameter estimated to be about 3440 miles (3550 kilometers). { 'tīt·ən }

Titania [ASTRON] A satellite of Uranus, with a diameter estimated to be 990 miles (1600 kilometers). { tī'tā·nē·ə }

Titius-Bode law *See* Bode's law. { 'tēt·sē·əs 'bōd·ə ,lȯ }

TLP *See* transient lunar phenomena.

topside sounder [AERO ENG] A satellite designed to measure ion concentration in the ionosphere from above the ionosphere. { 'täp,sīd ¦sau̇n·dər }

Toro [ASTRON] A small asteroid with a diameter of about 3.6 miles (6 kilometers), whose orbit, with semimajor axis of 1.368 astronomical units and eccentricity of 0.44, oscillates with that of Venus; it is about 0.13 astronomical unit from Earth at closest approach. { 'tȯr·ō }

total eclipse [ASTRON] An eclipse that obscures the entire surface of the moon or sun. { 'tōd·əl i'klips }

total impulse [AERO ENG] The product of the thrust and the time over which the thrust is produced, expressed in pounds (force)-seconds; used especially in reference to a rocket motor or a rocket engine. { 'tōd·əl 'im,pəls }

totality [ASTRON] **1.** The portion of a total eclipse of the sun during which the sun is entirely covered by the moon at a specified location on the earth's surface. **2.** The portion of a total eclipse of the moon or other body during which the eclipsed body is entirely within the umbra of the eclipsing body. { tō'tal·əd·ē }

Toucan *See* Tucana. { 'tü,kan }

tower telescope [ASTRON] A telescope, usually of long focal length, that is situated underneath a solar tower to study the sun. { 'taü·ər ¦tel·ə,skōp }

track [AERO ENG] The actual line of movement of an aircraft or a rocket over the surface of the earth; it is the projection of the history of the flight path on the surface. Also known as flight track. { trak }

trail [ASTRON] A luminous trace left in the sky by the passage of a large meteor. { trāl }

train [ASTRON] The bright tail of a comet or meteor. { trān }

transfer orbit [AERO ENG] In interplanetary travel, an elliptical trajectory tangent to the orbits of both the departure planet and the target planet. Also known as transfer ellipse. { 'tranz·fər ,ȯr·bət }

transient lunar phenomena [ASTRON] Local obscurations and reddish glows that are sometimes observed in certain areas of the moon. Abbreviated TLP. { 'tranch·ənt 'lü·nər fə'näm·ə·nä }

transient x-ray source *See* x-ray nova. { 'tranch·ənt 'eks,rā ,sȯrs }

transit [ASTRON] **1.** A celestial body's movement across the meridian of a place. Also known as meridian transit. **2.** Passage of a smaller celestial body across a larger one. **3.** Passage of a satellite's shadow across the disk of its primary. { 'trans·ət }

transition region [ASTRON] A layer of the solar atmosphere only a few hundred miles thick between the chromosphere and the corona across which the temperature rises rapidly from a few times 10^4 K to the order of 10^6 K. { tran'zish·ən ,rē·jən }

translunar [ASTRON] Beyond the orbit of the moon. { tran'slü·nər }

trap [AERO ENG] That part of a rocket motor that keeps the propellant grain in place. { trap }

Trapezium [ASTRON] Four very hot stars that appear to the eye as a single star in the Great Nebula of Orion; the star symbol is M42. { trə'pē·zē·əm }

Triangle *See* Triangulum. { 'trī,aŋ·gəl }

Triangulum [ASTRON] A northern constellation, right ascension 2 hours, declination 30°N. Also known as Triangle. { trī'aŋ·gyə·ləm }

Triangulum Australe [ASTRON] A southern constellation, right ascension 16 hours, declination 65°S. Also known as Southern Triangle. { trī'aŋ·gyə·ləm ȯ'strä·lē }

Triangulum Nebula [ASTRON] A nebula that is part of a small cluster of galaxies known as the local group; the nebula is labeled M 33. { trī'aŋ·gyə·ləm 'neb·yə·lə }

Trifid nebula [ASTRON] An emission nebula in Sagittarius that consists mostly of hydrogen ionized by hot, young stars, and displays dark lanes formed by dust. { 'trī,fid 'neb·yə·lə }

trigonometric parallax [ASTRON] A parallax that may be determined for the nearest stars (less than 300 light-years or 2.84×10^{18} m) by a direct method utilizing trigonometry. { ¦trig·ə·nə¦me·trik 'par·ə,laks }

triple-alpha process *See* three-alpha process. { 'trip·əl ¦al·fə ,prä·səs }

triple-mode Cepheid [ASTRON] A bent Cepheid that displays three nearly identical pulsation periods. { 'trip·əl ¦mōd 'sef·ē·əd }

Triton [ASTRON] The largest satellite of Neptune, with a diameter of about 1681 miles (2705 kilometers), orbiting at a mean distance of 220,500 miles (354,800 kilometers) with a period of 5 days 21.0 hours. { 'trīt·ən }

Trojan asteroid [ASTRON] **1.** One of a group of asteroids orbiting near the equilateral Lagranginan stability points of the sun-Jupiter system, which are located on Jupiter's orbit, 60° ahead of or 60° behind Jupiter. Also known as Jupiter Trojan. **2.** More generally, an object that is orbiting near one of the equilateral Lagrangian stability points of any pair of bodies. { ,trō·jən 'as·tə,rȯid }

tropical month [ASTRON] The average period of the revolution of the moon about the earth with respect to the vernal equinox, a period of 27 days 7 hours 43 minutes 4.7 seconds, or approximately $27\frac{1}{3}$ days. { 'träp·ə·kəl 'mənth }

tropical year [ASTRON] A unit of time equal to the period of one revolution of the earth about the sun measured between successive vernal equinoxes; it is 365.2422 mean solar days or 365 days 5 hours 48 minutes 46 seconds. Also known as astronomical year. { 'träp·ə·kəl 'yir }

Tropic of Cancer [ASTRON] A small circle on the celestial sphere connecting points with declination 23.45° north of the celestial equator, the northernmost declination of the sun. { 'träp·ik əv 'kan·sər }

Tropic of Capricorn [ASTRON] A small circle on the celestial sphere connecting points with declination 23.45° south of the celestial equator, the southernmost declination of the sun. { 'träp·ik əv 'kap·ri,körn }

true anomaly See anomaly. { 'trü ə'näm·ə·lē }

true place [ASTRON] The position of a star on the celestial sphere as it would be observed from the center of the sun, referred to the celestial equator and celestial equinox at the moment of observation. { 'trü 'plās }

true solar day See apparent solar day. { 'trü 'sō·lər 'dā }

true solar time See apparent solar time. { 'trü 'sō·lər 'tīm }

true sun See apparent sun. { 'trü 'sən }

T Tauri star [ASTRON] A star, with mass from 0.5 to 2.5 solar masses, in an early stage of formation at which interaction with its associated nebulosity, as well as possible internal instabilities, make it variable in luminosity and render its spectrum very peculiar. Also known as nebular variable. { 'tē 'tör·ē ,stär }

Tucana [ASTRON] A constellation in the southern hemisphere; right ascension 23 hours, declination 60° south. Also known as Toucan. { tü'kä·nə }

Tully-Fisher relation [ASTRON] A relation between the rotational velocity of a galaxy, as reflected in the width of the 21-centimeter line, and the intrinsic luminosity of the galaxy. { ¦təl·ē 'fish·ər ri,lā·shən }

tumbling [AERO ENG] An attitude situation in which the vehicle continues on its flight, but turns end over end about its center of mass. { 'təm·bliŋ }

Tundra orbit [AERO ENG] An inclined, elliptical, geosynchronous earth satellite orbit designed to provide communications satellite service coverage at high latitudes; it has an inclination of 63.4° degrees and an eccentricity of 0.2684, so that the satellite spends 16 hours each day over the hemisphere (northern or southern) where service coverage is intended and 8 hours over the opposite hemisphere. { 'tən·drə ,ór·bət }

turbidity [ASTRON] The formation of disks centered on stars in long-exposure photographs as a result of light scattering by grains in the emulsion. { tər'bid·əd·ē }

turnoff mass [ASTRON] The mass of those stars in a cluster that are at the turnoff point. { 'tərn,óf ,mas }

turnoff point [ASTRON] The point on a Hertzsprung-Russell diagram of a star cluster at which stars leave the main sequence and move toward the giant branch. { 'tərn,óf ,póint }

twilight [ASTRON] An intermediate period of illumination of the sky before sunrise and after sunset; the three forms are civil, nautical, and astronomical. { 'twī,līt }

twilight correction [ASTRON] In the interpretation of the records of sunshine recorders, the difference between the time of sunrise and the time at which a record of sunshine first began to be made by the sunshine recorder; and conversely at sunset; this correction is added only when the horizon is clear during the period. { 'twī,līt kə,rek·shən }

twilight zone [ASTRON] That zone of the earth or other planet in twilight at any time. { 'twī,līt ,zōn }

twinkling stars [ASTRON] Rapid fluctuations of the brightness and size of the images of stars caused by turbulence in the earth's atmosphere. { 'twink·liŋ 'stärz }

twin ring structures [ASTRON] Consanguineous ring structures whose component craters are of the same size, as well as being similar in form. { 'twin 'riŋ ,strək·chərz }

Twins See Gemini. { twinz }

Tycho [ASTRON] A crater on the near side of the moon. { 'tī·kō }

Tychonic system

Tychonic system [ASTRON] A theory of the planetary motion proposed by the astronomer Tycho Brahe in which the earth is stationary, with the sun and moon revolving about it but all the other planets revolving about the sun. { tī'kän·ik ˌsis·təm }

Tycho's Nova [ASTRON] A supernova that appeared in the constellation Cassiopeia in 1572; the star B Cassiopeiae. Also known as Tycho's star. { 'tī·kōz 'nō·və }

Tycho's star *See* Tycho's Nova. { 'tī·kōz 'stär }

Type II Cepheids *See* W Virginis stars. { 'tīp ˌtü 'sef·ē·ədz }

type I supernova [ASTRON] A member of a class of supernovae that lack hydrogen in their spectra and have relatively regular light curves. { 'tīp ˌwən ˈsü·pər'nō·və }

type Ia supernova [ASTRON] A member of a subclass of the type I supernovae that brighten relatively smoothly to a maximum whose brightness is relatively uniform among members of the subclass about 2 weeks after the explosion; decline in brightness quasi-exponentially thereafter; show strong lines of silicon, sulfur, calcium, and iron in their peak spectra; have late-time spectra dominated by strong emission lines of iron and cobalt; have line widths that imply velocities from 4000 to 12,000 miles (7000 to 20,000 kilometers) per second, with the highest velocities seen early on; and are believed to be exploding white dwarf stars. { ˌtīp ˌwən͵ā ˌsü·pər'nō·və }

type Ib supernova [ASTRON] A member of a subclass of the type I supernovae that lack a strong line feature at 615.0 nanometers due to ionized silicon; are approximately a factor of 4 fainter than type Ia and have light curves that are a little broader and slower to decline; have, in addition to iron emission, broad lines of oxygen and calcium in their late-time spectra; and are believed to be explosions of massive stars that are devoid of hydrogen. { ˌtīp ˌwən'bē ˈsü·pər͵nō·və }

type Ic supernova [ASTRON] A member of a subclass of the type I supernovae that resemble type Ib supernovae closely but have a weak or absent line of neutral helium at 587.6 nanometers. { ˌtīp ˌwən 'sē ˌsü·pər͵nō·və }

type II supernova [ASTRON] A member of a class of supernovae that display prominent lines of hydrogen in their spectra and have irregular light curves; they are believed be explosions of young massive stars, still in possession of their hydrogenic surface layers. { 'tīp ˌtü ˈsü·pər'nō·və }

type II-L supernova [ASTRON] A member of a subclass of the type II supernovae that begin to fade almost immediately in a quasi-linear way. { ˌtīp ˌtü 'el ˈsü·pər͵nō·və }

type II-P supernova [ASTRON] A member of a subclass of the type II supernovae that display enduring emission at a nearly constant rate for several months. { ˌtīp ˌtü 'pē ˈsü·pər͵nō·və }

U

UBV photometry [ASTRON] A system of three-color photometry used to obtain specific stellar magnitudes; the system is based on the comparison of stars' magnitudes with a standard sequence of about 400 stars. { ¦yü,bē¦vē fə'täm·ə·trē }

UBVRI system [ASTRON] An extension of the UBV system through measurements of an object's apparent magnitude with red (R) and near-infrared (I) filters. { ¦yü¦bē ¦vē¦är'ī ,sis·təm }

UBV system [ASTRON] A system of stellar magnitudes in which an object's apparent magnitude is measured at three wavelengths, labeled U, at 360 nanometers; B, at 420 nanometers; and V, at 540 nanometers; and is characterized by the color indices B-V and U-B, which are both defined to be 0 for a star of spectral type A0. Also known as Johnson-Morgan system. { ¦yü,bē¦vē ,sis·təm }

U Cephei [ASTRON] A binary star; in this double-star eclipsing system, one component has reached its Roche limit (a dynamical barrier beyond which the size of neither star can expand) while the other is distinctly smaller than this limit. { 'yü 'sef·ē,ī }

U Geminorum stars [ASTRON] A class of variable stars known as dwarf novae; their light curves resemble those of novae, with range brightness variations of about 4 magnitudes; examples are U Gemini and SS Cygni. Also known as SS Cygni stars. { 'yü ,jem·ə'nȯr·əm ,stärz }

ULIRG See ultraluminous infrared galaxy.

ullage rocket [AERO ENG] A small rocket used in space to impart an acceleration to a tank system to ensure that the liquid propellants collect in the tank in such a manner as to flow properly into the pumps or thrust chamber. { 'əl·ij ,räk·ət }

ultimate lines [ASTRON] Special spectral lines that can be used to indicate the existence of an element in the sun or other star. { 'əl·tə·mət 'līnz }

ultraluminous infrared galaxy [ASTRON] One of a class of galaxies that emit intense radiation at infrared wavelengths, 100 times or more as much as the Milky Way Galaxy. Abbreviated ULIRG. { ¦əl·trə¦lü·mə·nəs ,in·frə,red 'gal·ik·sē }

ultraviolet astronomy [ASTRON] Astronomical investigations utilizing observations carried out in the spectral region from approximately 350 to 90 nanometers. { ¦əl·trə'vī·lət ə'strän·ə·mē }

ultraviolet-bright star [ASTRON] A star that is brighter than stars on the horizontal branch and bluer than stars on the giant branch. { ¦əl·trə'vī·lət 'brīt 'stär }

ultraviolet star [ASTRON] A very hot star that is evolving toward the white dwarf stage; usually the central star of a planetary nebula. { ¦əl·trə'vī·lət 'stär }

umbilical connections [AERO ENG] Electrical and mechanical connections to a launch vehicle prior to lift off; the umbilical tower adjacent to the vehicle on the launch pad supports these connections which supply electrical power, control signals, data links, propellant loading, high pressure gas transfer, and air conditioning. { əm'bil·ə·kəl kə'nek·shənz }

umbilical cord [AERO ENG] Any of the servicing electrical or fluid lines between the ground or a tower and an uprighted rocket vehicle before the launch. Also known as umbilical. { əm'bil·ə·kəl ,kȯrd }

umbilical tower [AERO ENG] A vertical structure supporting the umbilical cords running into a rocket in launching position. { əm'bil·ə·kəl 'taů·ər }

umbra [ASTRON] The dark, central region of a sunspot. { 'əm·brə }

Umbriel [ASTRON] A satellite of Uranus orbiting at a mean distance of 165,300 miles (266,000 kilometers) with a period of 4.144 days, and with a diameter of about 740 miles (1190 kilometers). { 'əm·brē·əl }

universal time *See* Greenwich mean time. { ¦yü·nə¦vər·səl 'tīm }

universal time 0 [ASTRON] The uncorrected time of the earth's rotation as measured by the transit of stars across the observer's meridian. Abbreviated UT 0. { ¦yü·nə¦vər·səl ¦tīm 'zir·ō }

universal time 1 [ASTRON] Universal time 0 corrected for polar motion; it is the true angular rotation. Abbreviated UT 1. { ¦yü·nə¦vər·səl ¦tīm 'wən }

universal time 2 [ASTRON] Universal time 1 corrected for seasonal variations in the earth's rotation. Abbreviated UT 2. { ¦yü·nə¦vər·səl ¦tīm 'tü }

universal time coordinated [ASTRON] The coordinated time kept by a uniformly running clock, approximating the measure UT 2. Abbreviated UTC. { ¦yü·nə¦vər·səl ¦tīm kō'ȯrd·ən‚ād·əd }

universe [ASTRON] The totality of astronomical things, events, relations, and energies capable of being described objectively. { 'yü·nə‚vərs }

upper culmination *See* upper transit. { 'əp·ər ‚kəl·mə'nā·shən }

upper limb [ASTRON] That half of the outer edge of a celestial body having the greatest altitude. { 'əp·ər 'lim }

upper transit [ASTRON] The movement of a celestial body across a celestial meridian's upper branch. Also known as superior transit; upper culmination. { 'əp·ər 'trans·ət }

uranography [ASTRON] The science of mapping stars, groups of stars, and star clusters. { ‚yūr·ə'näg·rə·fē }

uranometry [ASTRON] The science of the measurement of the celestial sphere and celestial bodies. { ‚yūr·ə'näm·ə·trē }

Uranus [ASTRON] A planet, seventh in the order of distance from the sun; it has five known satellites, and its equatorial diameter is about four times that of the earth. { 'yūr·ə·nəs *or* yū'rā·nəs }

Urca process [ASTROPHYS] A series of nuclear reactions, chiefly among the iron group of elements, that are postulated as a cause of stellar collapse, due to the energy lost to neutrinos that are rapidly formed in the reactions. { 'ər·kə ‚prä‚ses }

Ursa Major [ASTRON] A northern constellation, right ascension 11 hours, declination 50°N; it contains a group of seven stars known as the Big Dipper. { 'ər·sə 'mā·jər }

Ursa Major cluster [ASTRON] **1.** An open cluster of stars, including 5 bright stars of the constellation Ursa Major, centered about 75 light-years from the sun and spread over a volume of 30 light-years length and 18-light-years width. **2.** A cluster of galaxies in the constellation Ursa Major having a redshift of about 0.051. { 'ər·sə 'mā·jər ‚kləs·tər }

Ursa Minor [ASTRON] A northern constellation, right ascension 15 hours, declination 70°N; its brightest star, Polaris, is almost at the north celestial pole; seven of the eight stars form a dipper ouline. Also known as Little Bear; Little Dipper. { 'ər·sə 'mī·nər }

Ursa Minor system [ASTRON] A dwarf spheroidal galaxy in the Local Group, about 2.1 × 10⁵ light-years (1.2 × 10¹⁸ miles or 2.0 × 10¹⁸ kilometers) away. { 'ər·sə 'mī·nər ‚sis·təm }

ursids [ASTRON] A shower of meteors occurring about December 22 from a radiant in the constellation Ursa Minor. { 'ər·sədz }

U Sagittae [ASTRON] An eclipsing binary star in which one component has attained its Roche limit, while its mate is distinctly smaller than this limit. { 'yü 'saj·ə‚tē }

UT 0 *See* universal time 0.

UT 1 *See* universal time 1.

UT 2 *See* universal time 2.

UTC *See* universal time coordinated.

uvby system [ASTRON] A four-color stellar magnitude system based on measurements in the ultraviolet, violet, blue, and yellow regions. Also known as Strömgren four-color index. { 'yüv·bē ‚sis·təm }

UV Ceti stars [ASTRON] A class of stars that have brief outbursts of energy over their surface areas; they may have an increase of about 1 magnitude for periods of 1 hour; the type star is UV Ceti. Also known as flare stars. { ¦yü¦vē 'sed·ē ˌstärz }

V

Valles Marineris [ASTRON] A system of canyons which extends for over 3000 miles (5000 kilometers) along the equatorial region of Mars; it is over 300 miles (500 kilometers) wide in places and drops to more than 4 miles (6 kilometers) below the surrounding surface. { 'val·əs ˌmar·ə'ner·is }

variable nebula [ASTRON] A nebula whose shape and brightness vary; an example is in the constellation Monoceros. { 'ver·ē·ə·bəl 'neb·yə·lə }

variable star [ASTRON] A star that has a detectable change in its intensity which is often accompanied by other physical changes; changes in brightness may be a few thousandths of a magnitude to 20 magnitudes or even more. { 'ver·ē·ə·bəl 'stär }

variational inequality [ASTRON] An inequality in the moon's motion, due mainly to the tangential component of the sun's attraction. { ˌver·ē·ā·shən·əl ˌin·i'kwäl·əd·ē }

vector steering [AERO ENG] A steering method for rockets and spacecraft wherein one or more thrust chambers are gimbal-mounted so that the direction of the thrust force (thrust vector) may be tilted in relation to the center of gravity of the vehicle to produce a turning movement. { 'vek·tər ˌstir·iŋ }

Vega [ASTRON] One of the brightest stars, apparent magnitude 0.1; it is a main sequence star of spectral type A0, distance is 8 parsecs, and it is 40 times brighter than the sun. Also known as α Lyrae. { 'vā·gə }

Vega-excess star [ASTRON] A star from which is detected far-infrared emission greatly in excess of what the star alone should produce, believed to originate from a cloud of dust surrounding the star that may be analogous to the Kuiper belt dust in the solar system but has hundreds or thousands of times as much dust. { 'vā·gə 'ek,ses ˌstär }

vehicle [AERO ENG] **1.** A structure, machine, or device, such as an aircraft or rocket, designed to carry a burden through air or space. **2.** More restrictively, a rocket vehicle. { 've·ə·kəl }

vehicle control system [AERO ENG] A system, incorporating control surfaces or other devices, which adjusts and maintains the altitude and heading, and sometimes speed, of a vehicle in accordance with signals received from a guidance system. Also known as flight control system. { 've·ə·kəl kən'trōl ˌsis·təm }

vehicle mass ratio [AERO ENG] The ratio of the final mass of a vehicle after all propellant has been used, to the initial mass. { 've·ə·kəl 'mas 'rā·shō }

Veil Nebula *See* Cygnus loop. { 'vāl 'neb·yə·lə }

Vela [ASTRON] A southern constellation, right ascension 9 hours, declination 50°S. Also known as Sail. { 'vē·lə }

Vela pulsar [ASTRON] A pulsar with a period of 80 milliseconds, about 1500 light-years (1.4×10^{19} meters) away in the constellation Vela, whose variation has been detected at radio, gamma-ray, and optical wavelengths; probably associated with the Vela supernova remnant. { 'vē·lə 'pəl,sär }

Vela supernova remnant [ASTRON] A gaseous nebula that is the result of a supernova whose light reached earth about 10,000 years ago. { 'vē·lə ¦sü·pər'nō·və ˌrem·nənt }

Vela X [ASTRON] A compact, nonthermal radio source associated with the Vela pulsar but displaced from it by about 0.7°. { 'vē·lə 'eks }

Vela X-1 [ASTRON] A pulsing, eclipsing x-ray source in the constellation Vela that is a particularly intense emitter of hard x-rays. { 'vē·lə ˌeks'wən }

Vela X-2 [ASTRON] The pulsed x-ray emission associated with the Vela pulsar. { 'vē·lə ,eks'tü }

velocity curve [ASTRON] A graphical representation of the line-of-sight velocity (versus time) of a star or components of a spectroscopic binary system. { və'läs·əd·ē ,kərv }

velocity-distance relation [ASTRON] The relation wherein all the exterior galaxies are moving away from the galaxy that the sun is part of, with velocities that are greater with increasing distance of the galaxy. { və'läs·əd·ē 'dis·təns ri,lā·shən }

velocity-of-light cylinder [ASTRON] A cylinder whose axis is the axis of rotation of a neutron star and whose radius is such that the velocity of a plasma rotating with the neutron star would equal the velocity of light at the surface of the cylinder. Also known as light cylinder. { və¦läs·əd·ē əv ¦līt 'sil·ən·dər }

velocity-of-light radius [ASTRON] The radius of the velocity-of-light cylinder. Also known as light radius. { və¦läs·əd·ē əv ¦līt 'rād·ē·əs }

Venus [ASTRON] The planet second in distance from the sun; the linear equatorial diameter of the solid globe is 7521 miles (12,104 kilometers); the mass is about 0.815 (earth = 1). { 'vē·nəs }

Venus probe [AERO ENG] A probe for exploring and reporting on conditions on or about the planet Venus, such as Pioneer and Mariner probes of the United States, and Venera probes of the Soviet Union. { 'vē·nəs 'prōb }

vernal equinox [ASTRON] The sun's position on the celestial sphere about March 21; at this time the sun's path on the ecliptic crosses the celestial equator. Also known as first point of Aries; March equinox; spring equinox. { 'vərn·əl 'ē·kwə,näks }

vernier engine [AERO ENG] A rocket engine of small thrust used primarily to obtain a fine adjustment in the velocity and trajectory of a rocket vehicle just after the thrust cutoff of the last sustainer engine, and used secondarily to add thrust to a booster or sustainer engine. Also known as vernier rocket. { 'vər·nē·ər 'en·jən }

vertex [ASTRON] **1.** The highest point that a celestial body attains. **2.** On a great circle, that point that is closest to a pole. { 'vər,teks }

vertical circle [ASTRON] A great circle of the celestial sphere, through the zenith and nadir of the celestial sphere; vertical circles are perpendicular to the horizon. { 'vərd·ə·kəl 'sər·kəl }

Very Large Array [ASTRON] An array near Socorro, New Mexico, of 27 separate radio telescopes on movable platforms, arranged along the arms of a Y, designed to provide radio pictures which have an angular resolution comparable with that of the best optical telescopes. Abbreviated VLA. { ¦ver·ē ¦lärj ə'rā }

Vesta [ASTRON] The third-largest asteroid with a diameter of about 300 miles (500 kilometers), mean distance from the sun of 2.362 astronomical units, and a unique surface composition resembling basaltic, achondritic meteorites. { 'ves·tə }

V471 Tauri star [ASTRON] A binary star, displaying irregular variability, in which one component is a main-sequence star and the other is either a white dwarf star or a star that is evolving toward the white dwarf state and is surrounded by a planetary nebula. { ¦vē,fȯr,sev·ən·tē,wən 'tȯr·ē ,stär }

violet layer [ASTRON] A layer of particles in the upper Martian atmosphere that scatter and absorb electromagnetic radiation at shorter wavelengths, making the atmosphere opaque to blue, violet, and ultraviolet light. { 'vī·ə·lət ,lā·ər }

Virgin *See* Virgo. { 'vər·jən }

Virgo [ASTRON] A constellation, right ascension 13 hours, declination 0°. Also known as Virgin. { 'vər·gō }

Virgo A [ASTRON] A radio galaxy; it is associated with the galaxy M 87 (NGC 4486). { 'vər·gō 'ā }

Virgo cluster [ASTRON] A cluster of galaxies which is the nearest to the galaxy that includes the sun; the cluster is centered in the constellation Virgo and is about 1.6 × 10^7 light-years (1.51 × 10^{23} m) from earth. { 'vər·gō ,kləs·tər }

Virgo supercluster *See* Local supercluster. { 'vər·gō 'sü·pər,kləs·tər }

Virgo X-1 [ASTRON] An x-ray source that is identical with Virgo A. { 'vər·gō ,eks'wən }

virial-theorem mass [ASTRON] The mass of a cluster of stars or galaxies calculated from the observed mean-square velocity of the objects and application of the virial theorem. { ¦vir·ē·əl 'thir·əm ˌmas }

visible horizon [ASTRON] That line where earth and sky appear to meet, and the projection of this line upon the celestial sphere. { 'viz·ə·bəl hə'rīz·ən }

visual binaries [ASTRON] Binary stars that to the naked eye seem to be single stars, but when viewed through the telescope, are separated into pairs. Also known as visual doubles. { 'vizh·ə·wəl 'bī‚ner·ēz }

visual doubles *See* visual binaries. { 'vizh·ə·wəl 'dəb·əlz }

visual magnitude [ASTRON] A celestial body's magnitude as seen by the eye of the observer. { 'vizh·ə·wəl 'mag·nəˌtüd }

VLA *See* Very Large Array.

Vogt-Russell theorem [ASTROPHYS] A theorem that states that if the pressure, opacity, and rate of energy generation in a star depend only on the local values of temperature, density, and chemical composition, then the star's structure is uniquely determined by its mass and chemical composition. Also known as Russell-Vogt theorem. { 'fōkt 'rəs·əl ˌthir·əm }

Volans [ASTRON] A southern constellation, right ascension 8 hours, declination 70°S. Also known as Flying Fish; Piscis Volans. { 'vōˌlanz }

volcanic theory [ASTRON] A theory which holds that most features of the moon's surface were formed by volcanic eruptions, lava flows, and subsidences when lunar rocks were plastic. Also known as igneous theory; plutonic theory. { väl'kan·ik 'thē·ə·rē }

Vulcan [ASTRON] A hypothetical planet that was supposed to have an orbit within the orbit of Mercury; its existence was considered about 1859 and in the next few years, but it is generally considered by present-day astronomers to be nonexistent. { 'vəl·kən }

Vulpecula [ASTRON] A northern constellation, right ascension 20 hours, declination 25°N. Also known as Little Fox. { ˌvəl'pek·yə·lə }

VV Cephei stars [ASTRON] A class of long-period eclipsing binary stars, with M supergiant primaries, a blue (usually B) supergiant or giant secondaries, and small variations in light. { ¦vē¦vē 'sef·ēˌī ˌstärz }

waning moon [ASTRON] The moon between full and new when its visible part is decreasing. { 'wān·iŋ 'mün }

Warrior *See* Orion. { 'wär·ē·ər }

Water Bearer *See* Aquarius. { 'wȯd·ər 'ber·ər }

Water Monster *See* Hydra. { 'wȯd·ər ‚män·stər }

Water Snake *See* Hydrus. { 'wȯd·ər ‚snāk }

waxing moon [ASTRON] The moon between new and full, when its visible part is increasing. { 'waks·iŋ 'mün }

weak-line T Tauri star [ASTRON] A T Tauri star that lacks strong emission lines in its optical spectrum, and lacks both strong stellar winds and a circumstellar accretion disk. Also known as naked T Tauri star; weak T Tauri star. { ¦wēk ‚līn ¦tē 'tȯr·ē ‚stär }

weak T Tauri star *See* weak-line T Tauri star. { ¦wēk ¦tē 'tȯr·ē ‚stär }

weathercocking [AERO ENG] The aerodynamic action causing alignment of the longitudinal axis of a rocket with the relative wind after launch. Also known as weather vaning. { 'weth·ər‚käk·iŋ }

week [ASTRON] A time period of 7 days which has been accepted from ancient Babylon; the 7 days of the week were first given names of the seven celestial bodies: the sun, moon, and five visible planets. { wēk }

Weizsaecker's theory [ASTRON] A theory of the origin of the solar system; it hypothesizes primeval turbulent eddies which become permanent and self-gravitating; Weizsaecker does not discuss the origin of the gas clouds. { 'vīt‚sek·ərz ‚thē·ə·rē }

west point [ASTRON] That point on the celestial sphere that is due west of observer; at this point the celestial equator crosses the horizon. { 'west 'pȯint }

wet emplacement [AERO ENG] A launch emplacement that provides a deluge of water for cooling the flame bucket, the rocket engines, and other equipment during the launch of a missile. { 'wet im'plās·mənt }

Whale *See* Cetus. { wāl }

Whirlpool galaxy [ASTRON] A spiral galaxy of type Sc (open spiral structure), seen face on, in the constellation Canes Venatici. { 'wərl‚pül ‚gal·ik·sē }

white dwarf star [ASTRON] An intrinsically faint star of very small radius and high density; the mass is about 0.6 that of the sun and the average radius is about 5000 miles (8000 kilometers); it is one final stage of stellar evolution with thermonuclear energy sources extinct. { 'wīt ¦dwȯrf 'stär }

Wilson-Bappu effect [ASTRON] A linear relation between the absolute magnitudes of late-type stars and the width of the K_2 emission core in the resonance line of ionized calcium (Ca II) at a wavelength of 3933 nanometers. { 'wil·sən 'bä·pü i‚fekt }

Wilson effect [ASTRON] An effect in which the penumbra of a sunspot appears narrower in the direction toward the sun's center than in the direction toward the sun's limb. { 'wil·sən i‚fekt }

window [AERO ENG] An interval of time during which conditions are favorable for launching a spacecraft on a specific mission. { 'win·dō }

Winged Horse *See* Pegasus. { 'wiŋd 'hȯrs }

winter [ASTRON] The period from the winter solstice, about December 22, to the vernal equinox, about March 21; popularly and for most meteorological purposes, winter is taken to include December, January, and February in the Northern Hemisphere, and June, July, and August in the Southern Hemisphere. { 'win·tər }

winter solstice |ASTRON| **1.** The sun's position on the ecliptic (about December 22). Also known as first point of Capricorn. **2.** The date (December 22) when the greatest southern declination of the sun occurs. { 'win·tər 'säl·stəs }

Wolf 359 |ASTRON| A star of absolute magnitude 16.6; it is 7.8 light-years from the sun and is a variable flare star, which may emit bursts of light and even radio noise. { wúlf ¦thrē¦fif·tē'nīn }

Wolf-Lundmark-Melotte galaxy |ASTRON| A dwarf irregular galaxy that is about 1.3 to 1.8 megaparsecs distant and is probably a member of the Local Group. { 'vólf 'lünd,märk mə'lät ¸gal·ik·sē }

Wolf number See relative sunspot number. { 'vólf ¸nəm·bər }

Wolf-Rayet nebula |ASTRON| A bright ring-shaped nebula that is radiatively ionized by a central Wolf-Rayet star. Also known as Wolf-Rayet ring. { ¦vólf rī¦ā ¸neb·yə·lə }

Wolf-Rayet ring See Wolf-Rayet nebula. { ¦vólf rī¦ā ¸riŋ }

Wolf-Rayet star |ASTRON| A member of a class of very hot stars (100,000–35,000 K) which characteristically show broad bright emission lines in their spectra; luminosities are high, probably in the range $10^4–10^5$ times that of the sun; these stars are probably very young and represent an early short-lived stage in stellar evolution. { ¦vólf rī¦ā ¸stär }

Wolf-Wolfer number See relative sunspot number. { 'vólf 'vúl·fər ¸nəm·bər }

Wolter type I x-ray telescope |ASTRON| A telescope that uses mirrors that form two surfaces of revolution, a paraboloid and a hyperboloid, to reflect incident x-rays to a common focus. { ¦wól·tər ¸tīp ¸wən ¦eks,rā 'tel·ə,skōp }

Workshop See Sculptor. { 'wərk,shäp }

world calendar |ASTRON| A proposed calendar in which the present 12 months are retained but the days are divided into four equal quarters; January, April, July, and October begin on Sunday and have 31 days, the other months have 30 days, so that there are 364 days with the 365th day following December 30 in no month; leap-year days would follow June 30. { 'wərld 'kal·ən·dər }

Wright's phenomenon |ASTRON| The phenomenon that the diameter of Mars appears greater when viewed from earth in ultraviolet light than when viewed in infrared light. { 'rīts fa,näm·ə,nän }

wrinkle ridge |ASTRON| A prominent, well-defined, often sinuous ridge on a lunar mare, with gently sloping sides and a height of generally less than 500 feet (150 meters). { 'riŋ·kəl ¸rij }

W stars |ASTRON| Stars of the W spectral class; their spectra contain an abundance of highly ionized elements such as He, C, N, and O, and they are intensely hot with surface temperatures of about 50,000 to 100,000 K. { 'dəb·əl·yü ¸stärz }

W Ursae Majoris stars |ASTRON| Eclipsing variable stars whose brightness is continuously varying in periods of a few hours; they are composed of two close stars that have a common gaseous envelope. { ¦dəb·əl·yü 'ər,sī mə'jór·əs ¸stärz }

W Virginis stars |ASTRON| Periodic variable stars with periods of about 10 to 30 days; they exhibit two surges of activity from the same star so that there is a doubling of their spectral lines. Also known as Type II Cepheids. { ¦dəb·əl·yü vər'jin·əs ¸stärz }

X

x-ray astronomy [ASTRON] The study of x-rays mainly from sources outside the solar system; it includes the study of novae and supernovae in the Milky Way Galaxy, together with extragalactic radio sources.   { 'eks ‚rā ə'strän·ə·mē }

x-ray background [ASTRON] Diffuse, almost isotropic x-radiation from beyond the solar system, believed to be the summed contribution of many unresolved sources.   { 'eks ‚rā 'bak‚graůnd }

x-ray binary [ASTRON] An x-ray source that is a member of a binary system.   { 'eks ‚rā 'bī‚ner·ē }

x-ray burster [ASTRON] One of a class of celestial x-ray sources which produce bursts of x-rays in the 1–20-kiloelectronvolt range and which are characterized by rise times of less than a few seconds and decay times of a few seconds to a few minutes; the peak luminosity is of the order of 10^{38} ergs per second (10^{31} watts) and the sources have an average equivalent temperature of 10^8 K.   { 'eks ‚rā 'bər·stər }

x-ray cluster [ASTRON] A cluster of galaxies that is pervaded by a diffuse medium that emits x-rays.   { 'eks ‚rā ‚kləs·tər }

x-ray nebulae [ASTRON] The remnant of an ancient supernova that has been identified as a source of x-rays; an example is the Crab Nebula.   { 'eks ‚rā 'neb·yə‚lī }

x-ray nova [ASTRON] An x-ray source which appears suddenly in the sky, dramatically increases in intensity over a few days, and then decays away with a lifetime of several months.   Also known as transient x-ray source.   { 'eks ‚rā 'nō·və }

x-ray star [ASTROPHYS] A source of x-rays from outside the solar system; examples are the point x-ray sources Scorpius X-1, Cygnus X-2, and the Crab x-ray source.   { 'eks ‚rā ‚stär }

Y

Yarkovsky effect [ASTRON] The effect of a small particle's rotation on its orbit about the sun, due to anisotropic reradiation. { yär'käf·skē i,fekt }

year [ASTRON] Any of several units of time based on the revolution of the earth about the sun; the tropical year to which the calendar is adjusted is the period required for the sun's longitude to increase 360°; it is about 365.24220 mean solar days. Abbreviated yr. { yir }

Yerkes system *See* MK system. { 'yər·kēz ,sis·təm }

ylem [ASTROPHYS] The primordial matter which according to the big bang theory existed just prior to the formation of the chemical elements. { 'ī·ləm }

yr *See* year.

Z

ZAMS *See* zero-age main sequence.

Z cam [ASTRON] A representative type of variable star; it is eruptive with a cycle of about 10–600 days; magnitude ranges from 2 to 6. { 'zē ,kam }

Z Camelopardalis stars [ASTRON] A class of dwarf novae which exhibit unpredictable, and sometimes very protracted, pauses in the decline from maximum to minimum brightness. { 'zē kə,mel·ə'pärd·əl·əs ,stärz }

zenith [ASTRON] That point of the celestial sphere vertically overhead. { 'zē·nəth }

zenithal hourly rate [ASTRON] The number of meteors in a meteor shower which would be observed per hour if the radiant of the meteor shower were overhead and there were no moonlight. { 'zē·nə·thəl 'au̇r·lē 'rāt }

zenith angle [ASTRON] The angle between the direction to the zenith and the direction of a light ray. { 'zē·nəth ,aŋ·gəl }

zenith distance [ASTRON] Angular distance from the zenith; the arc of a vertical circle between the zenith and a point on the celestial sphere, measured from the zenith through 90°, for bodies above the horizon. Also known as co-altitude. { 'zē·nəth ,dis·təns }

zenocentric coordinates [ASTRON] Coordinates that indicate the position of a point on the surface of Jupiter, determined by the direction of a line joining the center of Jupiter to the point. { ¦zēn·ə¦sen·trik kō'ȯrd·ən·əts }

zenographic coordinates [ASTRON] Coordinates that indicate the position of a point on the surface of Jupiter, determined by the direction of a line perpendicular to the mean surface at the point. { ¦zēn·ə¦graf·ik kō'ȯrd·ən·əts }

zero-age main sequence [ASTRON] The position on the Hertzsprung-Russell diagram of a star that has just reached the main sequence, when it has reached hydrostatic equilibrium and thermonuclear reactions have begun in its core, but these reactions have not had time to produce an appreciable change in composition. Abbreviated ZAMS. { 'zir·ō ¦āj 'mān 'sē·kwəns }

Zeta Aurigae star [ASTRON] A binary system with a supergiant primary in spectral class K and a main-sequence secondary. { 'zād·ə 'ȯr·ə,gē ,stär }

Zeta Geminorum stars [ASTRON] A subgroup of classical Cepheid variable stars whose variation of magnitude with time for one complete cycle produces a quasi-bell-shaped curve. { 'zād·ə ,jem·ə'nȯr·əm ,stärz }

Zodiac [ASTRON] A band of the sky extending 8° on each side of the ecliptic, within which the moon and principal planets remain. { 'zō·dē,ak }

zodiacal constellations [ASTRON] The constellations Aries, Taurus, Gemini, Cancer, Leo, Virgo, Libra, Scorpio, Sagittarius, Capricorn, Aquarius, and Pisces which are assigned to 12 equal portions of the zodiac. { zō'dī·ə·kəl ,kän·stə'lā·shənz }

zodiacal counterglow *See* gegenschein. { zō'dī·ə·kəl 'kau̇nt·ər,glō }

zodiacal dust [ASTRON] A cloud of dust that fills the plane of the solar system interior to the asteroid belt, and is responsible for zodiacal light. { zō'dī·ə·kəl 'dəst }

zone meridian [ASTRON] The meridian used for reckoning zone time; this is generally the nearest meridian whose longitude is exactly divisible by 15°. { 'zōn mə'rid·ē·ən }

zone noon [ASTRON] Twelve o'clock zone time, or the instant the mean sun is over the upper branch of the zone meridian. { 'zōn 'nün }

zone of avoidance [ASTRON] An irregularly shaped area in the Milky Way Galaxy in which no extragalactic nebulae are observed because of the presence of interstellar matter. { 'zōn əv ə'vȯid·əns }

zone time [ASTRON] The local mean time of a reference or zone meridian whose time is kept throughout a designated zone; the zone meridian is usually the nearest meridian whose longitude is exactly divisible by 15°. { 'zōn 'tīm }

Z time *See* Greenwich mean time. { 'zē ,tīm }

Zulu time *See* Greenwich mean time. { 'zü·lü ,tīm }

Zurich number *See* relative sunspot number. { 'zur·ik ,nəm·bər }

Zwicky dark matter [ASTRON] Matter of unknown nature that is postulated to exist outside the visible parts of galaxies or between galaxies in order to account for the radial velocity dispersion of galaxies in clusters. { ¦zwik·ē 'därk ,mad·ər }

ZZ Ceti stars [ASTRON] A small class of variable-luminosity white dwarfs with small amplitude oscillations having periods of 10–100 seconds and effective temperatures of 10,000–14,000 K (18,000–25,000°F). { ¦zē¦zē 'sed·ē ,stärz }

Appendix

Equivalents of commonly used units for the U.S. Customary System and the metric system

1 inch = 2.5 centimeters (25 millimeters)	1 centimeter = 0.4 inch	1 inch = 0.083 foot
1 foot = 0.3 meter (30 centimeters)	1 meter = 3.3 feet	1 foot = 0.33 yard (12 inches)
1 yard = 0.9 meter	1 meter = 1.1 yards	1 yard = 3 feet (36 inches)
1 mile = 1.6 kilometers	1 kilometer = 0.62 mile	1 mile = 5280 feet (1760 yards)
1 acre = 0.4 hectare	1 hectare = 2.47 acres	
1 acre = 4047 square meters	1 square meter = 0.00025 acre	
1 gallon = 3.8 liters	1 liter = 1.06 quarts = 0.26 gallon	1 quart = 0.25 gallon (32 ounces; 2 pints)
1 fluid ounce = 29.6 milliliters	1 milliliter = 0.034 fluid ounce	1 pint = 0.125 gallon (16 ounces)
32 fluid ounces = 946.4 milliliters		1 gallon = 4 quarts (8 pints)
1 quart = 0.95 liter	1 gram = 0.035 ounce	1 ounce = 0.0625 pound
1 ounce = 28.35 grams	1 kilogram = 2.2 pounds	1 pound = 16 ounces
1 pound = 0.45 kilogram	1 kilogram = 1.1×10^{-3} ton	1 ton = 2000 pounds
1 ton = 907.18 kilograms		
°F = (1.8 × °C) + 32	°C = (°F − 32) ÷ 1.8	

Conversion factors for the U.S. Customary System, metric system, and International System

A. Units of length

Units	cm	m	in.	ft	yd	mi
1 cm =	1	0.01	0.3937008	0.03280840	0.01093613	6.213712×10^{-6}
1 m =	100.	1	39.37008	3.280840	1.093613	6.213712×10^{-4}
1 in. =	2.54	0.0254	1	0.08333333...	0.02777777...	1.578283×10^{-5}
1 ft =	30.48	0.3048	12.	1	0.3333333...	$1.893939... \times 10^{-4}$
1 yd =	91.44	0.9144	36.	3.	1	$5.681818... \times 10^{-4}$
1 mi =	1.609344×10^{5}	1.609344×10^{3}	6.336×10^{4}	5280.	1760.	1

B. Units of area

Units	cm²	m²	in.²	ft²	yd²	mi²
1 cm² =	1	10^{-4}	0.1550003	1.076391×10^{-3}	1.195990×10^{-4}	3.861022×10^{-11}
1 m² =	10^{4}	1	1550.003	10.76391	1.195990	3.861022×10^{-7}
1 in.² =	6.4516	6.4516×10^{-4}	1	$6.944444... \times 10^{-3}$	7.716049×10^{-4}	2.490977×10^{-10}
1 ft² =	929.0304	0.09290304	144.	1	0.1111111...	3.587007×10^{-8}
1 yd² =	8361.273	0.8361273	1296.	9.	1	3.228306×10^{-7}
1 mi² =	2.589988×10^{10}	2.589988×10^{6}	4.014490×10^{9}	2.78784×10^{7}	3.0976×10^{6}	1

C. Units of volume

Units	m^3	cm^3	liter	$in.^3$	ft^3	qt	gal
1 m^3 =	1	10^6	10^3	6.102374×10^4	35.31467	1.056688×10^3	264.1721
1 cm^3 =	10^{-6}	1	10^{-3}	0.06102374	3.531467×10^{-5}	1.056688×10^{-3}	2.641721×10^{-4}
1 liter =	10^{-3}	1000.	1	61.02374	0.03531467	1.056688	0.2641721
1 $in.^3$ =	1.638706×10^{-5}	16.38706	0.01638706	1	5.787037×10^{-4}	0.01731602	4.329004×10^{-3}
1 ft^3 =	2.831685×10^{-2}	28316.85	28.31685	1728.	1	29.92208	7.480520
1 qt =	9.463529×10^{-4}	946.3529	0.9463529	57.75	0.03342014	1	0.25
1 gal (U.S.) =	3.785412×10^{-3}	3785.412	3.785412	231.	0.1336806	4.	1

D. Units of mass

Units	g	kg	oz	lb	metric ton	ton
1 g =	1	10^{-3}	0.03527396	2.204623×10^{-3}	10^{-6}	1.102311×10^{-6}
1 kg =	1000.	1	35.27396	2.204623	10^{-3}	1.102311×10^{-3}
1 oz (avdp) =	28.34952	0.02834952	1	0.0625	2.834952×10^{-5}	3.125×10^{-5}
1 lb (avdp) =	453.5924	0.4535924	16.	1	4.535924×10^{-4}	$5. \times 10^{-4}$
1 metric ton =	10^8	1000.	35273.96	2204.623	1	1.102311
1 ton =	907184.7	907.1847	32000.	2000.	0.9071847	1

Appendix

Conversion factors for the U.S. Customary System, metric system, and International System (cont.)

E. Units of density

Units	$g \cdot cm^{-3}$	$g \cdot L^{-1}, kg \cdot m^{-3}$	$oz \cdot in.^{-3}$	$lb \cdot in.^{-3}$	$lb \cdot ft^{-3}$	$lb \cdot gal^{-1}$
$1\ g \cdot cm^{-3}$ =	1	1000.	0.5780365	0.03612728	62.42795	8.345403
$1\ g \cdot L^{-1}, kg \cdot m^{-3}$ =	10^{-3}	1	5.780365×10^{-4}	3.612728×10^{-5}	0.06242795	8.345403×10^{-3}
$1\ oz \cdot in.^{-3}$ =	1.729994	1729.994	1	0.0625	108.	14.4375
$1\ lb \cdot in.^{-3}$ =	27.67991	27679.91	16.	1	1728.	231.
$1\ lb \cdot ft^{-3}$ =	0.01601847	16.01847	9.259259×10^{-3}	5.787037×10^{-4}	1	0.1336806
$1\ lb \cdot gal^{-1}$ =	0.1198264	119.8264	4.749536×10^{-3}	4.329004×10^{-3}	7.480519	1

F. Units of pressure

Units	$Pa, N \cdot m^{-2}$	$dyn \cdot cm^{-2}$	bar	atm	$kgf \cdot cm^{-2}$	$mmHg\ (torr)$	$in.\ Hg$	$lbf \cdot in.^{-2}$
$1\ Pa, 1\ N \cdot m^{-2}$ =	1	10	10^{-5}	9.869233×10^{-6}	1.019716×10^{-5}	7.500617×10^{-3}	2.952999×10^{-4}	1.450377×10^{-4}
$1\ dyn \cdot cm^{-2}$ =	0.1	1	10^{-6}	9.869233×10^{-7}	1.019716×10^{-6}	7.500617×10^{-4}	2.952999×10^{-5}	1.450377×10^{-5}
$1\ bar$ =	10^5	10^6	1	0.9869233	1.019716	750.0617	29.52999	14.50377
$1\ atm$ =	101325	1013250	1.01325	1	1.033227	760.	29.92126	14.69595
$1\ kgf \cdot cm^{-2}$ =	98066.5	980665	0.980665	0.9678411	1	735.5592	28.95903	14.22334
$1\ mmHg\ (torr)$ =	133.3224	1333.224	1.333224×10^3	1.315789×10^{-3}	1.359510×10^{-3}	1	0.03937008	0.01933678
$1\ in.\ Hg$ =	3386.388	33863.88	0.03386388	0.03342105	0.03453155	25.4	1	0.4911541
$1\ lbf \cdot in.^{-2}$ =	6894.757	68947.57	0.06894757	0.06804596	0.07030696	51.71493	2.036021	1

G. Units of energy

Units	g mass (energy equiv.)	J	eV	cal	cal_{IT}	Btu_{IT}	kWh	hp-h	ft-lbf	$ft^3 \cdot lbf \cdot in.^{-2}$	liter-atm
1 g mass (energy equiv.) = 1	1	8.987552×10^{13}	5.609589×10^{32}	2.148076×10^{13}	2.146640×10^{13}	8.518555×10^{10}	2.496542×10^{7}	3.347918×10^{7}	6.628878×10^{13}	4.603388×10^{11}	8.870024×10^{11}
1 J =	1.112650×10^{-14}	1	6.241510×10^{18}	0.2390057	0.2388459	9.478172×10^{-4}	$2.777777... \times 10^{-7}$	3.725062×10^{-7}	0.7375622	5.121960×10^{-3}	9.869233×10^{-3}
1 eV =	1.782662×10^{-33}	1.602176×10^{-19}	1	3.829293×10^{-20}	3.826733×10^{-20}	1.518570×10^{-22}	4.450490×10^{-26}	5.968206×10^{-26}	1.181705×10^{-19}	8.206283×10^{-22}	1.581225×10^{-21}
1 cal =	4.655328×10^{-14}	4.184	2.611448×10^{19}	1	0.9993312	3.965667×10^{-3}	$1.1622222... \times 10^{-6}$	1.558562×10^{-6}	3.085960	2.143028×10^{-2}	0.04129287
1 cal_{IT} =	4.658443×10^{-14}	4.1868	2.613195×10^{19}	1.000669	1	3.968321×10^{-3}	1.163×10^{-6}	1.559609×10^{-6}	3.088025	2.144462×10^{-2}	0.04132050
1 Btu_{IT} =	1.173908×10^{-11}	1055.056	6.585141×10^{21}	252.1644	251.9958	1	2.930711×10^{-4}	3.930148×10^{-4}	778.1693	5.403953	10.41259
1 kWh =	4.005540×10^{-8}	3600000.	2.246944×10^{25}	860420.7	859845.2	3412.142	1	1.341022	2655224.	18349.06	35529.24
1 hp-h =	2.986931×10^{-8}	2384519.	1.675545×10^{25}	641615.6	641186.5	2544.33	0.7456998	1	1980000.	13750.	26494.15
1 ft-lbf =	1.508551×10^{-14}	1.355818	8.462351×10^{18}	0.3240483	0.3238315	1.285067×10^{-3}	3.766161×10^{-7}	$5.050505... \times 10^{-7}$	1	$6.944444... \times 10^{-3}$	0.01338088
1 $ft^3 \cdot lbf \cdot in.^{-2}$ =	2.172313×10^{-12}	195.2378	1.218579×10^{21}	46.66295.	46.63174	0.1850497	5.423272×10^{-5}	$7.272727... \times 10^{-5}$	144.	1	1.926847
1 liter-atm =	1.127393×10^{-12}	101.325	6.324210×10^{20}	24.21726	24.20106	0.09603757	2.814583×10^{-5}	3.774419×10^{-5}	74.73349	0.5189825	1

Appendix

Total solar eclipses, 1998–2020*

Date	Maximum duration of totality	Location
Feb. 26, 1998	4 m 08 s	Pacific Ocean, north of South America, Caribbean islands, Atlantic Ocean
Aug. 11, 1999	2 m 23 s	Atlantic Ocean, Europe, southwestern Asia, India
June 21, 2001	4 m 56 s	Atlantic Ocean, South Africa, Madagascar
Dec. 4, 2002	2 m 04 s	South Africa, Indian Ocean, Australia
Nov. 23, 2003	1 m 57 s	Antarctica
Apr. 8, 2005	0 m 42 s†	Pacific Ocean
Mar. 29, 2006	4 m 07 s	Atlantic Ocean, Africa, Turkey, Central Asia
Aug. 1, 2008	2 m 27 s	Northern Canada, Greenland, Siberia, China
July 22, 2009	6 m 40 s	India, China, Pacific Ocean
July 11, 2010	5 m 20 s	Pacific Ocean, south of South America
Nov. 13, 2012	4 m 02 s	Australia, Pacific Ocean
Nov. 3, 2013	1 m 40 s†	Atlantic Ocean, Africa
Mar. 20, 2015	2 m 47 s	Arctic (including Spitsbergen)
Mar. 9, 2016	4 m 10 s	Indonesia
Aug. 21, 2017	2 m 40 s	United States (Oregon to Georgia)
July 2, 2019	4 m 33 s	Pacific Ocean, Chile, Argentina
Dec. 4, 2020	2 m 10 s	Pacific Ocean, Chile, Argentina, Atlantic

*Compiled by Jean Meeus.
†These eclipses are annular-total; that is, they are total only near the middle of their central line.

Lunar eclipses in the umbra, 1999–2019

Date	Magnitude	Date	Magnitude
July 28, 1999	0.40	Dec. 31, 2009	0.08
Jan. 21, 2000	1.33	June 26, 2010	0.54
July 16, 2000	1.77	Dec. 21, 2010	1.26
Jan. 9, 2001	1.19	June 15, 2011	1.70
July 5, 2001	0.50	Dec. 10, 2011	1.11
May 16, 2003	1.13	June 4, 2012	0.37
Nov. 9, 2003	1.02	Apr. 25, 2013	0.02
May 4, 2004	1.31	Apr. 15, 2014	1.29
Oct. 28, 2004	1.31	Oct. 8, 2014	1.17
Oct. 17, 2005	0.07	Apr. 4, 2015	1.00
Sept. 7, 2006	0.19	Sept. 28, 2015	1.28
Mar. 3, 2007	1.24	Aug. 7, 2017	0.25
Aug. 28, 2007	1.48	Jan. 31, 2018	1.32
Feb. 21, 2008	1.11	July 27, 2018	1.61
Aug. 16, 2008	0.81	Jan. 21, 2019	1.20

Physical characteristics of the Sun's planets

Planet	Equatorial radius (r_e)			Ellipticity	Volume (Earth = 1)	Mass (Earth = 1)	Density, g/cm³	Escape velocity		Rotation period	Obliquity, degrees[1]
	(Earth = 1)	mi	km					mi/s	km/s		
Mercury	0.38	1,515	2,440	0.000	0.055	0.055	5.43	2.5	4.4	58 d 15.5 h	0.1
Venus	0.95	3,761	6,052	0.000	0.854	0.815	5.20	6.5	10.4	243 d 0.5 h	177.4[2]
Earth	1.00	3,963	6,378	0.0034	1.000	1.000	5.52	7.0	11.2	23 h 56 m 23 s	23.45
Mars	0.53	2,110	3,396	0.0069	0.151	0.107	3.34	3.1	5.0	24 h 37 m 23 s	25.19
Jupiter	11.21	44,423	71,492	0.0649	1408.	317.710	1.33	37.0	59.5	9 h 55 m 30 s[3,4]	3.12
Saturn	9.45	37,449	60,268	0.0980	844.	95.162	0.69	22.1	35.5	10 h 39 m 22 s[3,5]	26.73
Uranus	4.01	15,882	25,559	0.0229	64.	14.535	1.32	13.2	21.3	17 h 22.2 m[3,6]	97.86[2]
Neptune	3.88	15,389	24,766	0.017	59.	17.141	1.64	14.6	23.5	16 h 6.6 m[3,7]	29.56
Pluto	0.18	715	1,150	?	0.006	0.002	2.0	0.7	1.1	6 d 9 h 17.6 m	119.6[2]

[1] Obliquity is the tilt of the equator with respect to the orbit plane.
[2] Venus, Uranus, and Pluto are considered to have retrograde rotation.
[3] Internal (System III) rotation period, the rotation period of the planet's core, as deduced from its magnetic field.
[4] Jupiter's equatorial (System I) rotation period is 9 h 50.5 m.
[5] Saturn's equatorial rotation period is 10 h 14.0m.
[6] Uranus's equatorial rotation period is about 18.0 h.
[7] Neptune's equatorial rotation period is about 18.8 h.

Appendix

Elements of planetary orbits

Planet	Symbol	Mean distance from Sun (semimajor axis of orbit)			Sidereal period of revolution		Synodic period, days	Mean orbital velocity		Orbital eccentricity	Orbital inclination, degrees
		AU	10^6 mi	10^6 km	Years	Days		mi/s	km/s		
Mercury	☿	0.387	36.0	57.9	0.241	87.97	115.88	29.75	47.87	0.206	7.00
Venus	♀	0.723	67.2	108.2	0.615	224.70	583.92	21.76	35.02	0.007	3.39
Earth	⊕	1.000	93.0	149.6	1.000	365.24		18.51	29.79	0.017	0.00
Mars	♂	1.524	141.6	227.9	1.881	686.93	779.94	14.99	24.13	0.093	1.85
Jupiter	♃	5.203	483.6	778.3	11.857	4,330.60	398.88	8.12	13.07	0.048	1.30
Saturn	♄	9.555	888.2	1429.4	29.424	10,746.9	378.09	6.01	9.67	0.056	2.49
Uranus	♅	19.22	1786.	2875.	83.75	30,588.7	369.66	4.24	6.83	0.046	0.77
Neptune	♆	30.11	2799.	4504.	164.72	59,799.9	367.49	3.41	5.48	0.009	1.77
Pluto	♇	39.54	3676.	5916.	248.0	90,589.	366.72	2.95	4.75	0.249	17.14

Satellites of the planets*

Satellite	Mean distance from center of planet		Sidereal period, days	Mean diameter†	
	10^3 km	10^3 mi		10^3 km	10^3 mi
EARTH					
Moon	384.4	238.9	27.322	3,475	2,159
MARS					
Phobos	9.38	5.83	0.319	22	14
				$(26 \times 22 \times 19)$	$(16 \times 14 \times 12)$
Deimos	23.46	14.58	1.262	12	8
				$(16 \times 12 \times 10)$	$(10 \times 8 \times 6)$
JUPITER					
Metis	128.0	79.5	0.295	43 (60×34)	27 (37×21)
Adrastea	129.0	80.2	0.298	16 (20×14)	10 (12×9)
Amalthea	181.4	112.7	0.498	167	104
				(250×120)	(155×80)
Thebe	221.9	137.9	0.675	99 (116×84)	61 (72×52)
Io	421.8	262.1	1.769	3,643	2,264
Europa	671.1	417.0	3.551	3,122	1,940
Ganymede	1,070.4	665.1	7.155	5,262	3,270
Callisto	1,882.7	1,169.9	16.69	4,821	2,995
Themisto	7,507	4,665	130	8	5
Leda	11,165	6,938	241	20	12
Himalia	11,461	7,122	251	170	106
Lysithea	11,717	7,281	259	36	22
Elara	11,741	7,296	260	86	53
S/2000 J11	12,555	7,801	287	4	2.5
S/2001 J10	19,394	12,051	541	4	2.5
Iocaste	20,216	12,562	585	5	3
Praxidike	20,964	13,026	617	7	4
S/2001 J7	21,027	13,066	602	6	4
Harpalyke	21,132	13,131	625	4	3
S/2001 J9	21,168	13,153	604	4	2.5
S/2001 J3	21,252	13,205	613	8	5
Ananke	21,276	13,220	610	28	17
S/2001 J2	21,312	13,243	613	8	5
S/2001 J6	23,029	14,310	686	4	2.5
Isonoe	23,078	14,340	719	4	2
S/2001 J8	23,124	14,369	690	4	2.5
Erinome	23,168	14,396	723	3	2
S/2001 J4	23,219	14,428	691	6	4
Taygete	23,312	14,485	730	5	3
Chaldene	23,387	14,532	734	4	2
Carme	23,404	14,543	702	46	29
S/2001 J11	23,547	14,631	708	6	4
Pasiphae	23,624	14,679	708	60	37
Kalyke	23,745	14,754	751	5	3
S/2001 J5	23,808	14,794	715	4	2.5
Magaclite	23,911	14,858	758	5	3
Sinope	23,939	14,875	725	38	24
Callirrhoe	24,100	14,975	759	9	5
S/2001 J1	24,122	14,989	729	8	5

Appendix

Satellites of the planets* (cont.)

Satellite	Mean distance from center of planet 10^3 km	Mean distance from center of planet 10^3 mi	Sidereal period, days	Mean diameter† 10^3 km	Mean diameter† 10^3 mi
SATURN					
Pan	133.6	83.0	0.575	20 ± 6	12 ± 4
Atlas	137.7	85.6	0.602	32 ± 8	20 ± 5
Prometheus	139.4	86.6	0.613	100 ± 6	62 ± 4
Pandora	141.7	88.0	0.629	84 ± 4	52 ± 2
Epimetheus	151.4	94.1	0.694	119 ± 6	74 ± 4
Janus	151.5	94.1	0.695	178 ± 8	110 ± 5
Mimas	185.6	115.3	0.942	397	247
Enceladus	238.1	147.9	1.370	499	310
Tethys	294.7	183.1	1.888	1,060 ± 3	658 ± 2
Telesto	294.7	183.1	1.888	24 ± 6	15 ± 4
Calypso	294.7	183.1	1.888	19 ± 3	12 ± 2
Dione	377.4	234.5	2.737	1,118 ± 10	695 ± 6
Helene	377.4	234.5	2.737	32 ± 8	20 ± 5
Rhea	527.1	234.5	4.518	1,528 ± 8	949 ± 5
Titan	1,221.9	759.3	15.95	5,150 ± 4	3,200 ± 2
Hyperion	1,464.1	909.7	21.28	283 ± 40	176 ± 25
Iapetus	3,560.8	2,212.6	79.33	1,436 ± 16	892 ± 10
S/2000 S5	11,365	7,062	449	14	9
S/2000 S6	11,440	7,108	451	10	6
Phoebe	12,944.3	8,043.2	548	220 ± 20	137 ± 12
S/2000 S2	15,199	9,444	687	19	12
S/2000 S8	15,645	9,721	729	6	4
S/2000 S11	16,392	10,186	783	26	16
S/2000 S10	17,611	10,943	871	9	5
S/2000 S3	18,160	11,284	893	32	20
S/2000 S4	18,239	11,333	926	13	8
S/2000 S9	18,709	11,625	951	6	3.5
S/2000 S12	19,470	12,098	1,017	6	3.5
S/2000 S7	20,470	12,719	1,089	6	3.5
S/2000 S1	23,096	14,351	1,312	16	10
URANUS					
Cordelia	49.8	30.9	0.335	40 ± 6	25 ± 4
Ophelia	53.8	33.4	0.376	43 ± 8	27 ± 5
Bianca	59.2	36.8	0.435	51 ± 4	32 ± 2
Cressida	61.8	38.4	0.464	80 ± 4	49 ± 2
Desdemona	62.7	39.0	0.474	64 ± 8	40 ± 5
Juliet	64.4	40.0	0.493	94 ± 8	58 ± 5
Portia	66.1	41.1	0.513	135 ± 8	84 ± 5
Rosalind	69.9	43.4	0.558	72 ± 12	45 ± 7
Belinda	75.3	46.8	0.624	81 ± 16	50 ± 10
Puck	86.0	53.4	0.762	162 ± 4	101 ± 2
Miranda	129.9	80.7	1.413	472	293
Ariel	190.9	118.6	2.520	1,158	719
Umbriel	266.0	165.3	4.144	1,169 ± 6	727 ± 4
Titania	436.3	271.1	8.706	1,578 ± 4	980 ± 2
Oberon	583.5	362.6	13.46	1,523 ± 5	946 ± 3
Caliban	7,231	4,493	579	98	61
Stephano	8,004	4,973	677	20	12
S/2001 U1	8,578	5,330	758	7–19	4–12
Sycorax	12,179	7,568	1,283	190	118
Prospero	16,243	10,093	1,962	30	19
Setebos	17,501	10,875	2,209	30	19

Satellites of the planets *(cont.)*

Satellite	Mean distance from center of planet		Sidereal period, days	Mean diameter[†]	
	10^3 km	10^3 mi		10^3 km	10^3 mi
NEPTUNE					
Naiad	48.2	30.0	0.294	58 ± 12	36 ± 7
Thalassa	50.1	31.1	0.311	80 ± 16	50 ± 10
Despina	52.5	32.6	0.335	148 ± 20	92 ± 12
Galatea	62.0	38.5	0.429	158 ± 24	98 ± 15
Larissa	73.5	45.7	0.555	192 ± 14	119 ± 9
Proteus	117.6	73.1	1.122	416 ± 16	258 ± 10
Triton	354.8	220.5	5.877	2,707	1,682
Nereid	5,513.4	3,425.9	360.14	340 ± 50	211 ± 31
PLUTO					
Charon	19.41	12.06	6.387	1,186 ± 26	737 ± 16

[*]The orbits of the unnamed satellites of Jupiter and Saturn, discovered in 2000 and 2001, are still uncertain. Moreover, there are unresolved issues in defining parameters for the orbits of the outer satellites of Jupiter, since these orbits are strongly perturbed by the Sun and change significantly over long time periods. For these reasons, the values determined by different groups for these satellites disagree in many cases, and will probably be modified by further observations and analysis.

[†]Dimensions are also given for some significantly nonspherical satellites.

SOURCE: Solar System Dynamics Group, Jet Propulsion Laboratory.

Appendix

Noteworthy asteroids

Size rank	Number and name	Spectral type	Diameter mi (km)*	Spin period, h	Orbital elements		
					a. AU	e	i, degrees
1	1 Ceres	G (C-like)	578 (930)	9.1	2.77	0.08	10.6
2	2 Pallas	B (C-like)	343 (552)	7.8	2.77	0.23	34.8
3	4 Vesta	Achondrite	324 (521)	5.3	2.36	0.09	7.1
4	10 Hygiea	C	260 (419)	27.6	3.14	0.12	3.8
5	704 Interamnia	F (C-like)	203 (327)	8.7	3.06	0.15	17.3
6	511 Davida	C	200 (322)	5.1	3.18	0.17	15.9
7	52 Europa	C	183 (295)	5.6	3.09	0.11	7.5
8	87 Sylvia	P	172 (277)	5.2	3.48	0.09	10.9
9	65 Cybele	P	167 (269)	4.0	3.43	0.11	3.6
10	15 Eunomia	S	161 (259)	6.1	2.64	0.19	11.8
11	16 Psyche	M	155 (249)	4.2	2.92	0.14	3.1
12	31 Euphrosyne	B (C-like)	154 (248)	5.5	3.15	0.23	26.3
13	451 Patientia	B (C-like)	153 (247)	9.7	3.06	0.07	15.2
14	3 Juno	S	150 (242)	7.2	2.67	0.25	13.0
15	324 Bamberga	C	149 (240)	29.4	2.68	0.34	11.2
16	13 Egeria	G (C-like)	139 (224)	7.0	2.58	0.09	16.5
17	624 Hektor	D	186 × 93 (300 × 150)	6.9	5.15	0.03	18.3
18	532 Herculina	S	137 (220)	9.4	2.77	0.17	16.3
19	107 Camila	C	137 (220)	4.8	3.49	0.07	10.0
20	423 Diotima	P?	135 (217)	4.8	3.07	0.03	11.2
21	121 Hennione	C	135 (217)	9.2	3.46	0.14	7.6
22	45 Eugenia	F (C-like)	133 (215)	5.7	2.72	0.08	6.6
23	19 Fortuna	C	130 (210)	7.5	2.44	0.16	1.6
24	24 Themis	C	129 (207)	8.4	3.13	0.13	0.8
25	7 Iris	S	127 (204)	7.1	2.39	0.23	5.5
26	6 Hebe	S	126 (202)	7.3	2.42	0.20	14.8
27	702 Alauda	C	126 (202)	8.4	3.19	0.03	20.5
28	88 Thisbe	C	124 (200)	6.0	2.77	0.16	5.2

Number and name		Spectral type	Diameter mi (km)*	Spin period, h	Orbital elements		
					a, AU	e	i, degrees
Other interesting asteroids							
41	Daphne	C	116 (187)	6.0	2.77	0.27	15.8
44	Nysa	Aubrite?	42 (68)	6.4	2.42	0.15	3.7
165	Loreley	C	99 (160)	7.2	3.14	0.07	11.2
216	Kleopatra	M	87 (140)	5.4	2.79	0.25	13.2
243	Ida	S	19 (31)	4.6	2.86	0.04	1.1
250	Bettina	M	53 (86)	5.1	3.14	0.14	12.9
349	Dembowska	Achondrite?	90 (145)	4.7	2.93	0.09	8.3
433	Eros	Chon./S?	20 × 8 (33 × 13)	5.3	1.46	0.22	10.8
747	Winchester	P	116 (186)	9.4	3.00	0.34	18.2
951	Gaspra	S	7.6 (12)	7.0	2.21	0.17	4.1
1566	Icarus	Chondrite	1 (2)	2.3	1.08	0.83	22.9
1620	Geographos	S	1.3 (2.1)	5.2	1.24	0.33	13.3
4179	Toutatis	S	~2 (4)	130	2.51	0.64	0.5

*Diameters are accurate to about 10%; size rankings may vary.

Appendix

Major meteor showers

Shower	Duration	Approx. date of maximum	Approx. radiant coordinates, degrees*		Meteoroid orbital speed		Strength†	Suggested parent body	Notes
			Right ascension	Declination	mi/s	km/s			
Quadrantids	Jan. 1–6	Jan. 3	230	+49	27	42	M	—	Sharp maximum
Lyrids	Apr. 20–23	Apr. 22	271	+34	30	48	M–W	1861 I	Good in 1982
π Puppids	Apr. 16–25	Apr. 23	110	–45	—	—	W–M	Grigg-Skjellerup	Highly variable
η Aquarids	Apr. 21–May 12	May 4	336	–2	40	64	M	Halley	Second peak May 6
Arietids	May 29–June 19	June 7	44	+23	24	39	S	—	Daytime shower
ζ Perseids	June 1–17	June 7	62	+23	18	29	S	—	Daytime shower
β Taurids	June 24–July 6	June 29	86	+19	20	32	S	Encke	Daytime Shower
S. δ Aquarids	July 21–Aug. 25	July 30	333	–16	27	43	M	—	Primary radiant
S. ι Aquarids	July 15–Aug. 25	Aug. 6	333	–15	19	31	W	—	Primary radiant
Perseids	July 23–Aug. 23	Aug. 12	46	+57	37	60	S	1862 III	Best-known shower
Orionids	Oct. 2–Nov. 7	Oct. 21	94	+16	41	66	M	Halley	Trains common
S. Taurids	Sept. 15–Nov. 26	Nov. 3	50	+14	17	27	M	Encke	Known fireball producer
Leonids	Nov. 14–20	Nov. 17	152	+22	45	72	W–S	1866 I	Peak in 1999–2001
Puppids-Velids	Nov. 27–Jan.	Dec. 9	135	–48	—	—	M	2102 Tantalus?	Many radiants in region
Geminids	Dec. 4–16	Dec. 13	112	+32	23	36	S	3200 Phaethon	Many bright meteors

* Radiant coordinates are those of the apparent radiant rather than the geocentric radiant.
† Estimate of relative meteor hourly rate for visual observers. S = strong (sometimes above 30 per hour at peak); M = moderate (10 to 30 per hour at peak); W = weak (5 to 10 per hour at peak).

Minor meteor showers*

Shower	Duration	Approx. date of maximum	Approx. radiant coordinates, degrees		Meteoroid orbital speed		Suggested parent body	Notes
			Right ascension	Declination	mi/s	km/s		
Coma Berenicids	Dec. 12–Jan. 23	Jan. 17	186	+20	40	65	1913 I	Uncertain radiant position
α Centaurids	Jan. 28–Feb. 23	Feb. 8	209	−59	—	—	—	Colors in bright meteors
δ Leonids	Feb. 5–Mar. 19	Feb. 26	159	+19	14	23	—	Slow, bright meteors
Virginids	Feb. 3–Apr. 15	Mar. 13?	186	+00	21	35	—	Other radiants in region
δ Normids	Feb. 25–Mar. 22	Mar. 14	245	−49	—	—	—	Sharp maximum
δ Pavonids	Mar. 11–Apr. 16	Apr. 6	305	−63	—	—	Grigg-Mellish	Rich in bright meteors
σ Leonids	Mar. 21–May 13	Apr. 17	195	−05	12	20	—	Slow, bright meteors
α Scorpids	Apr. 11–May 12	May 3	240	−22	21	35	—	Other radiants in region
τ Herculids	May 19–June 14	May 3	228	+39	9	15	—	Very slow meteors
Ophiuchids	May 19–July 2	June 10	270	−23	—	—	—	One of many in region
Corvids	June 25–30	June 26	192	−19	6	11	—	Very low speed
June Draconids	June 5–July 19?	June 28	219	+49	8	14	Pons-Winnecke	Maximum only 1916
Capricornids	July–Aug.	July 8	311	−15	—	—	—	May be multiple
Piscis-Australids	July 15–Aug. 20	July 31	340	−30	—	—	—	Poorly known
α Capricornids	July 15–Aug. 25	Aug. 2	307	−10	14	23	—	Bright meteors
N. α Aquarids	July 14–Aug. 25	Aug. 12	327	−06	26	42	—	Secondary radiant
κ Cygnids	Aug. 9–Oct. 6	Aug. 18	286	+59	15	25	1948 XII	Bursts of activity
N. ι Aquarids	July 15–Sept. 20	Aug. 20	327	−06	19	31	—	Secondary radiant
S. Piscids	Aug. 31–Nov. 2	Sept. 20	6	+00	16	26	—	Primary radiant
Andromedids	Sept. 25–Nov. 12	Oct. 3	20	+34	11	18	—	"Annual" version
October Draconids	Oct. 10	Oct. 10	262	+54	14	23	Giacobini-Zinner	Can be spectacular
N. Piscids	Sept. 25–Oct. 19	Oct. 12	26	+14	18	29	—	Secondary radiant
Leo Minorids	Oct. 22–24	Oct. 24	162	+37	38	62	1739	Probable comet association
μ Pegasids	Oct. 29–Nov. 12	Nov. 12	335	+21	7	11	1819 IV	Probable comet association
Andromedids	Nov. 25	Nov. 25	25	+44	10	17	Biela	Once only, 1885
Phoenicids	Dec. 5	Dec. 5	15	−50	—	—	—	Once only, 1956
Ursids	Dec. 17–24	Dec. 22	217	+76	20	33	—	Good in 1986

*Peak strength for visual observers usually less than 5 per hour. Radiant coordinates are those of the apparent radiant rather than the geocentric radiant.

Appendix

The constellations*

Latin name	Genitive	Abbreviation	English translation
Andromeda	Andromedae	And	Andromeda[†]
Antlia	Antliae	Ant	Pump
Apus	Apodis	Aps	Bird of Paradise
Aquarius	Aquarii	Aqr	Water Bearer
Aquila	Aquilae	Aql	Eagle
Ara	Arae	Ara	Altar
Aries	Arietis	Ari	Ram
Auriga	Aurigae	Aur	Charioteer
Boötes	Boötis	Boo	Herdsman
Caelum	Caeli	Cae	Chisel
Camelopardalis	Camelopardalis	Cam	Giraffe
Cancer	Cancri	Cnc	Crab
Canes Venatici	Canum Venaticorum	CVn	Hunting Dogs
Canis Major	Canis Majoris	CMa	Big Dog
Canis Minor	Canis Minoris	CMi	Little Dog
Capricornus	Capricorni	Cap	Goat
Carina	Carinae	Car	Ship's Keel[‡]
Cassiopeia	Cassiopeiae	Cas	Cassiopeia[†]
Centaurus	Centauri	Cen	Centaur[†]
Cepheus	Cephei	Cep	Cepheus[†]
Cetus	Ceti	Cet	Whale
Chamaeleon	Chamaeleonis	Cha	Chameleon
Circinus	Circini	Cir	Compass
Columba	Columbae	Col	Dove
Coma Berenices	Comae Berenices	Com	Berenice's Hair[†]
Corona Australis	Coronae Australis	CrA	Southern Crown
Corona Borealis	Coronae Borealis	CrB	Northern Crown
Corvus	Corvi	Crv	Crow
Crater	Crateris	Crt	Cup
Crux	Crucis	Cru	Southern Cross
Cygnus	Cygni	Cyg	Swan
Delphinus	Delphini	Del	Dolphin
Dorado	Doradus	Dor	Swordfish
Draco	Draconis	Dra	Dragon
Equuleus	Equulei	Equ	Little Horse
Eridanus	Eridani	Eri	River Eridanus[†]
Fornax	Fornacis	For	Furnace
Gemini	Geminorum	Gem	Twins
Grus	Gruis	Gru	Crane
Hercules	Herculis	Her	Hercules[†]
Horologium	Horologii	Hor	Clock
Hydra	Hydrae	Hya	Hydra[†] (water monster)
Hydrus	Hydri	Hyi	Sea Serpent
Indus	Indi	Ind	Indian
Lacerta	Lacertae	Lac	Lizard
Leo	Leonis	Leo	Lion
Leo Minor	Leonis Minoris	LMi	Little Lion
Lepus	Leporis	Lep	Hare
Libra	Librae	Lib	Scales
Lupus	Lupi	Lup	Wolf
Lynx	Lyncis	Lyn	Lynx
Lyra	Lyrae	Lyr	Harp
Mensa	Mensae	Men	Table (mountain)
Microscopium	Microscopii	Mic	Microscope
Monoceros	Monocerotis	Mon	Unicorn

The constellations* *(cont.)*

Latin name	Genitive	Abbreviation	English translation
Musca	Muscae	Mus	Fly
Norma	Normae	Nor	Level (square)
Octans	Octantis	Oct	Octant
Ophiuchus	Ophiuchi	Oph	Ophiuchus[†] (serpent bearer)
Orion	Orionis	Ori	Orion[†]
Pavo	Pavonis	Pav	Peacock
Pegasus	Pegasi	Peg	Pegasus[†] (winged horse)
Perseus	Persei	Per	Perseus[†]
Phoenix	Phoenicis	Phe	Phoenix
Pictor	Pictoris	Pic	Easel
Pisces	Piscium	Psc	Fish
Piscis Austrinus	Piscis Austrini	PsA	Southern Fish
Puppis	Puppis	Pup	Ship's Stern[‡]
Pyxis	Pyxidis	Pyx	Ship's Compass[‡]
Reticulum	Reticuli	Ret	Net
Sagitta	Sagittae	Sge	Arrow
Sagittarius	Sagittarii	Sgr	Archer
Scorpius	Scorpii	Sco	Scorpion
Sculptor	Sculptoris	Scl	Sculptor
Scutum	Scuti	Sct	Shield
Serpens	Serpentis	Ser	Serpent
Sextans	Sextantis	Sex	Sextant
Taurus	Tauri	Tau	Bull
Telescopium	Telescopii	Tel	Telescope
Triangulum	Trianguli	Tri	Triangle
Triangulum Australe	Trianguli Australis	TrA	Southern Triangle
Tucana	Tucanae	Tuc	Toucan
Ursa Major	Ursae Majoris	UMa	Big Bear
Ursa Minor	Ursae Minoris	UMi	Little Bear
Vela	Velorum	Vel	Ship's Sails[‡]
Virgo	Virginis	Vir	Virgin
Volans	Volantis	Vol	Flying Fish
Vulpecula	Vulpeculae	Vul	Little Fox

*After J. M. Pasachoff, *Astronomy: From the Earth to the Universe*, 5th ed., 1998.
[†]Proper names.
[‡]Formerly formed the consellation Argo Navis, the Argonauts' Ship.

First-magnitude stars*

Proper name	Greek-letter name	Apparent visual magnitude (V)	Distance, light-years	Absolute visual magnitude (M_v)	Spectral class
Sirius	α Canis Majoris	−1.47	8.6	1.42	A1 V
Canopus	α Carinae	−0.72	315	−5.6	F0 II
Rigil Kentaurus	α Centauri	−0.29	4.4	4.06	G2 V + K1 V
Arcturus	α Boötis	−0.04	37	−0.31	K1.5 III
Vega	α Lyrae	0.03	25	0.58	A0 V
Capella	α Aurigae	0.08	42	−0.48	G5 III + G0 III
Rigel	β Orionis	0.12	780	−7	B8 Ia
Procyon	α Canis Minoris	0.34	11	2.67	F5 IV
Achernar	α Eridani	0.46	145	−2.76	B3 V
Betelgeuse	α Orionis	0.50	425	−5.1	M1.5 Iab
Hadar	β Centauri	0.60	525	−5.4	B1 III
Altair	α Aquilae	0.77	17	2.25	A7 V
Aldebaran	α Tauri	0.85	65	−0.66	K5 III
Acrux	α Crucis	0.94	320	−4.0	B0.5 IV + B1 V
Spica	α Virginis	1.04	260	−3.49	B2 V + B2 V
Antares	α Scorpii	1.09	600	−5.2	M1.5 Ib
Pollux	β Geminorum	1.14	34	1.08	K0 III
Fomalhaut	α Piscis Austrini	1.16	25	1.73	A3 V
Deneb	α Cygni	1.25	1600	−7	A2 Ia
Mimosa	β Crucis	1.30	350	−3.9	B0.5 III
Regulus	α Leonis	1.35	80	−0.55	B7 V

*As seen by eye, doubles combined into one. Doubles are indicated by two entries in the spectral class. Low-luminosity companions are not included.

Stars within 10 light-years*

Star	Distance, light-years	Spectral class	Apparent visual magnitude (V)	Absolute visual magnitude (M_v)
Proxima Centauri	4.22	M5.5 V	11.01	15.45
Alpha Centauri A	4.40	G2 V	−0.01	4.34
Alpha Centauri B	4.40	K1 V	1.33	5.68
Barnard's Star	5.94	M5 V	9.54	13.24
Wolf 359	7.79	M6.5 V	13.46	16.57
BD + 36° 2147	8.31	M2 V	7.49	10.46
Sirius A	8.60	A1 V	−1.47	1.42
Sirius B	8.60	DA[†]	8.44	11.33
L726-8A	8.7	M5.5 V	12.56	15.42
L726-8B	8.7	M5.5 V	12.96	15.82
Ross 154	9.69	M3.6 V	10.37	13.00

*Adapted from R. Bishop (ed.), *Observer's Handbook, 1999,* Royal Astronomical Society of Canada, University of Toronto Press, 1998.
[†]Hydrogen-rich white dwarf.

Spectral classes of stars

Class	Characteristic absorption lines	Color*	Temperature, K[†]
O	H, He+, He	Blue	30,000–55,000
B	H, He	Blue-white	9500–30,000
A	H	White	7000–9500
F	Metals, H	Yellow-white	6000–7000
G	Ca+, metals	Yellow	5200–6000
K	Ca+, Ca	Orange	3900–5200
M	TiO, other molecules, Ca	Orange-red	2000–3900
L[‡]	Metal hydrides	Red	<2100
T[¶]	Methane bands	Infrared	1000
R[§]	Carbon	Orange	4000–5800
N[§]	Carbon molecules	Red	2000–4000
S[§]	ZrO and other molecules	Orange-red	2000–4000

*The color refers to energy distribution; visual colors are subtle.
[†]Main sequence.
[‡]A mixture of red dwarf stars and brown dwarfs.
[¶]Brown dwarfs.
[§]R and N are combined into carbon stars, class C. Classes R, N, and S all consist of giants.

Appendix

Members of the Local Group of galaxies, ordered from brightest to faintest

Name	Type	Distance, 10^6 light-years	Solar luminosities
1. M31 = Andromeda	Spiral	2.5	3×10^{10}
2. Milky Way Galaxy	Spiral	—	2×10^{10}
3. M33 = Triangulum	Spiral	2.7	6×10^{9}
4. Large Magellanic Cloud	Irregular	0.16	3×10^{9}
5. IC 10	Irregular	2.7	9×10^{8}
6. Small Magellanic Cloud	Irregular	0.19	7×10^{8}
7. NGC 3109*	Irregular	4.4	3×10^{8}
8. NGC 205	Elliptical	2.5	3×10^{8}
9. NGC 6822	Irregular	1.7	2×10^{8}
10. WLM	Irregular	2.9	2×10^{8}
11. M32	Elliptical	2.5	2×10^{8}
12. NGC 185	Elliptical	2.1	2×10^{8}
13. NGC 404*	Elliptical	8.	2×10^{8}
14. NGC 147	Elliptical	2.2	1×10^{8}
15. IC 1613	Irregular	2.5	1×10^{8}
16. Sextans A*	Irregular	4.8	1×10^{8}
17. Sextans B*	Dwarf irregular	4.5	6×10^{7}
18. Pegasus	Dwarf irregular	3.1	3×10^{7}
19. DDO 210	Dwarf irregular	2.	3×10^{7}
20. Leo A*	Irregular	2.3	2×10^{7}
21. Sagittarius	Dwarf spheroidal	0.08	2×10^{7}
22. Antlia*	Dwarf irregular	4.3	1×10^{7}
23. Fornax	Dwarf spheroidal	0.46	7×10^{6}
24. Andromeda I	Dwarf spheroidal	2.5	6×10^{6}
25. Leo I	Dwarf spheroidal	0.89	5×10^{6}
26. DDO 155 = GR8*	Dwarf irregular	5.	5×10^{6}
27. DDO 187*	Dwarf irregular	5.	4×10^{6}
28. Andromeda II	Dwarf spheroidal	1.9	2×10^{6}
29. Sculptor	Dwarf spheroidal	0.25	2×10^{6}
30. Phoenix	Dwarf irregular	1.4	2×10^{6}
31. Andromeda III	Dwarf spheroidal	2.5	1×10^{6}
32. SagDIG	Dwarf irregular	1.6	1×10^{6}
33. Sextans	Dwarf spheroidal	0.26	1×10^{6}
34. Andromeda IV	Dwarf spheroidal	2.5	9×10^{5}
35. LGS 3	Dwarf irregular	2.6	8×10^{5}
36. Tucana	Dwarf spheroidal	2.8	7×10^{5}
37. Leo II	Dwarf spheroidal	0.70	6×10^{5}
38. Ursa Minor	Dwarf spheroidal	0.21	4×10^{5}
39. Carina	Dwarf spheroidal	0.28	4×10^{5}
40. Draco	Dwarf spheroidal	0.25	3×10^{5}

*Doubtful members.

Large optical telescopes

Meters	Inches	Observatory	Year completed
		Some of the world's largest reflecting telescopes	
10.4	410	Gran Telescopio Canarias, La Palma, Canary Islands	2003
10.2	400	Keck I Telescope, Mauna Kea, Hawaii	1993
10.2	400	Keck II Telescope, Mauna Kea, Hawaii	1996
11.1 × 9.8	437 × 385	Hobby-Eberly Telescope, McDonald Observatory, Fort Davis, Texas	1997
11.1 × 9.8	437 × 385	South African Large Telescope, Sutherland, South Africa	2004
8.4 (twin)	330 (twin)	Large Binocular Telescope, Mt. Graham, Arizona*	2003
8.3	327	Subaru Telescope, National Astronomical Observatory, Japan; Mauna Kea, Hawaii	2000
8.2	320	Antu (Very Large Telescope 1), Cerro Paranal, Chile†	1998
8.2	320	Kueyen (Very Large Telescope 2), Cerro Paranal, Chile†	1999
8.2	320	Melipal (Very Large Telescope 3), Cerro Paranal, Chile†	2000
8.2	320	Yepun (Very Large Telescope 4), Cerro Paranal, Chile†	2000
8.1	319	Gemini North, National Optical Astronomy Observatories, Mauna Kea, Hawaii	1999
8.1	319	Gemini South, National Optical Astronomy Observatories, Cerro Pachon, Chile	2000
6.5	255	MMT Observatory, Mount Hopkins, Arizona	1999
6.5	255	Walter Baade Telescope, Las Campanas, Chile	2000
6.5	255	Landon Clay Telescope, Las Campanas, Chile	2002
6.1	240	Large Zenith Telescope, Vancouver, Canada	2002
6.0	236	Special Astrophysical Observatory, Zelenchukskaya, Caucasus, Russia	1976
5.1	200	Hale Telescope, California Institute of Technology/ Palomar Observatory, Palomar Mountain, California	1950
4.2	165	William Herschel Telescope, La Palma, Canary Islands	1987
4.0	158	Mayall Telescope, Kitt Peak National Observatory, Arizona	1973
4.0	158	Victor Blanco Telescope, Cerro Tololo Inter-American Observatory, Chile	1976
3.9	153	Anglo-Australian Telescope, Siding Spring Observatory, Australia	1975
3.8	150	United Kingdom Infrared Telescope, Mauna Kea Observatory, Hawaii	1978
3.7	147	Advanced Electro-Optical System (AEOS), Maui, Hawaii	2000
3.6	144	Canada-France-Hawaii Telescope, Mauna Kea Observatory, Hawaii	1979
3.6	142	Cerro La Silla European Southern Observatory, Chile	1976
3.6	141	New Technology Telescope, European Southern Observatory, Chile	1989
3.6	141	Telescopio Nazionale Galileo, La Palma, Canary Islands	1998
3.5	138	Calar Alto Observatory, Calar Alto, Spain	1984
3.5	138	Astrophysics Research Consortium, Apache Point, New Mexico	1993
3.5	138	Wisconsin-Indiana-Yale-NOAO (WIYN), Kitt Peak, Arizona	1994
3.5	138	Starfire, Kirtland Air Force Base, New Mexico	1994
3.2	126	NASA Infrared Telescope, Mauna Kea Observatory, Hawaii	1979
3.0	120	Lick Observatory, Mount Hamilton, California	1959
2.7	107	McDonald Observatory, Fort Davis, Texas	1968
2.6	102	Crimean Astrophysical Observatory, Ukraine	1960

Appendix

Large optical telescopes (*cont.*)

Mirror diameter		Observatory	Year completed
Meters	Inches		
		Some of the world's largest reflecting telescopes (cont.)	
2.6	102	Shajn Telescope, Byurakan Astrophysical Observatory, Yerevan, Armenia	1976
2.6	101	Nordic Optical Telescope, La Palma, Canary Islands	1989
2.5	100	Hooker Telescope, Mount Wilson and Las Campanas, Mount Wilson, California	1917
2.5	100	de Pont Telescope, Cerro las Campanas, Carnegie Southern Observatory, Chile	1976
2.5	100	Sloan Digital Sky Survey, Apache Point, New Mexico	2000
2.5	96	Isaac Newton Telescope, La Palma, Canary Islands	1984
2.4	94	University of Michigan-Dartmouth College-Ohio State University-Columbia University, Kitt Peak, Arizona	1986
2.4	94	Hubble Space Telescope, low Earth orbit	1990
		World's largest refracting telescopes	
1.02	40	Yerkes Observatory, Williams Bay, Wisconsin	1897
0.91	36	Lick Observatory, Mount Hamilton, California	1888
0.83	33	Observatoire de Paris, Meudon, France	1893
0.80	32	Astrophysikalisches Observatory, Potsdam, Germany	1899
0.76	30	Allegheny Observatory, Pittsburgh, Pennsylvania	1914

*Two 8.4-m (330-in.) mirrors on a common mount.
†The four unit telescopes of the Very Large Telescope can operate independently or as a single instrument.

Largest Schmidt telescopes

Telescope or institution	Location	Diameter		Focal ratio	Date completed
		Corrector plate, in. (m)	Spherical mirror, in. (m)		
Karl Schwarzschild Observatory	Tautenberg (near Jena), Germany	53 (1.3) (removable)	79 (2.0)	f/2	1960
Oschin Telescope, Palomar Observatory	Palomar Mountain, California	48 (1.2)	72 (1.8)	f/2.5	1948, 1987
United Kingdom, Schmidt Anglo-Australian Observatory	Siding Spring Observatory, New South Wales, Australia	48 (1.2)	72 (1.8)	f/2.5	1973
Tokyo Astronomical Observatory	Kiso Mountains, Japan	41 (1.1)	60 (1.5)	f/3.1	1975
European Southern Observatory Schmidt*	La Silla, Chile	39 (1.0)	64 (1.6)	f/3	1972

*Decommissioned in 1998.

177

Appendix

Large radio telescopes and synthesis arrays

Institution	Location	Size of reflector, ft (m)
Radio telescopes for meter and centimeter wavelengths		
Fully steerable paraboloids		
Robert C. Byrd Green Bank Telescope	Green Bank, West Virginia	360 × 330 (110 × 100)
Max-Planck-Institut für Radioastronomie	Effelsberg, Germany	330 (100)
Nuffield Radio Astronomy Laboratory	Jodrell Bank, England	250 (76)
Tidbinbilla	Tidbinbilla, ACT, Australia	230 (70)
CSIRO	Parkes, N.S.W., Australia	211 (64)
Jet Propulsion Laboratory	Goldstone, California	211 (64)
Algonquin Radio Observatory	Lake Traverse, Ontario	152 (46)
National Radio Astronomy Observatory	Green Bank, West Virginia	142 (43)
California Institute of Technology	Big Pine, California	132 (40)
Haystack Observatory	Westford, Massachusetts	122 (37)
Crimean Astrophysical Observatory	Crimea, Russia	73 (22)
Limited-tracking transit telescopes		
RATAN-600 Special Astrophysical Observatory	Zelenchukskaya, Russia	33 × 6221 (10 × 1885)
Tata Institute	Ootacamund, India	99 × 1746 (30 × 529)
National Astronomy and Ionosphere Center	Arecibo, Puerto Rico	1007 (305)
Observatory of Paris	Nançay, France	132 × 660 (40 × 200)
Radio telescopes for millimeter wavelengths		
Nobeyama Radio Observatory*	Nobeyama, Japan	148 (45)
Instiut de RadioAstronomie Millimétrique	Pico Veleta, Spain	99 (30)
Onsala Observatory	Gothenburg, Sweden	66 (20)
University of Massachusetts	Amherst, Massachusetts	46 (14)
National Radio Astronomy Observatory	Kitt Peak, Arizona	40 (12)
California Institute of Technology†	Big Pine, California	33 (10)
University of Texas	Fort Davis, Texas	16(5)
University of California‡	Hat Creek, California	13 (4)
Synthesis arrays		
National Radio Astronomy Observatory (Very Long Baseline Array)	Socorro, New Mexico; Pie Town, New Mexico; Virgin Islands; Hawaii	Resolution 0.001″
Atacama Large Millimeter Array (ALMA)	Llano de Chajnantor, Chile	Resolution 0.01″
Australia Telescope National Facility	Narrabri, Coonabarabran, and Parkes, Australia	Resolution 0.01″
National Radio Astronomy Observatory (Very Large Array, VLA)	Socorro, New Mexico	Resolution 0.1″
Nuffield Radio Astronomy Laboratory (MERLIN)	Jodrell Bank, England	Resolution 0.1″
Institut de Radio Astronomie Millimétrique	Plateau de Bure, France	Resolution 0.1″
Mullard Radio Astronomy Observatory (5-km array)	Cambridge, England	Resolution 0.5″
Westerbork Radio Observatory (WSRT)	Westerbork, Netherlands	Resolution 1″
Giant Metrewave Radio Telescope (GMRT)	Pune, India	Resolution 2″

*Also five-element millimeter-wave interferometer.
†Also three-element millimeter-wave interferometer.
‡Four-element millimeter-wave interferometer.

Some telescopes for submillimeter astronomy

Name	Location	Elevation, ft (m)	Diameter, ft (m)
Existing installations			
Cologne 3-m Telescope	Gornergrat, near Zermatt, Switzerland	10,285 (3,135)	9.8 (3.0)
Caltech Submillimeter Observatory (CSO)	Mauna Kea, Hawaii	13,360 (4,072)	34.1 (10.4)
Swedish-ESO Submillimeter Telescope (SEST)	La Silla, Chile	7,850 (2,400)	49.2 (15.0)
James Clerk Maxwell Telescope (JCMT)	Mauna Kea, Hawaii	13,425 (4,092)	49.2 (15.0)
Max-Planck-Institut für Radioastronomie/University of Arizona 10-m Submillimeter Telescope	Mount Graham, Arizona	10,466 (3,190)	32.8 (10.0)
Smithsonian Sub-Millimeter Array (SMA)	Mauna Kea, Hawaii	13,385 (4,080)	19.7 (6.0) [8 telescopes]
Antarctic Submillimeter Telescope/Remote Observatory (AST/RO)	South Pole	9,340 (2,847)	5.6 (1.7)
National Radio Observatory of Japan	Mt. Fuji, Japan	12,218 (3,724)	4.0 (1.2)
Atacama Submillimeter Telescope Experiment (ASTE)	Pampa la Bola, Chile	15,750 (4,800)	32.8 (10.0)
Planned installations			
Stratospheric Observatory for Infrared Astronomy (SOFIA)*	Suborbital	50,000 (15,200)	9.8 (3.0)
Far-Infrared and Submillimeter Telescope (FIRST)†	Orbital	>200 mi (>320 km)	23.0 (7.0)
Large Deployable Reflector (LDR)‡	Orbital	>200 mi (>320 km)	32.8 (10.0)
South Pole Telescope¶	South Pole	9,340 (2,847)	26.4 (8.0)
Atacama Large Millimeter Array (ALMA)§	Llano de Chajnantor, Chile	16,400 (5,000)	39.4 (12.0) [at least 64 telescopes]

*Possible completion in 2004 (United States). Optical, infrared, and submillimeter.
†Possible deployment in 2006 (European).
‡Possible deployment about 2003 (United States). Operating wavelength 30 μm to 1 mm.
¶Scheduled to become operational in 2006 (United States).
§Early science operations scheduled for 2007 and full science operations for 2012.

Appendix

X-ray telescopes

Mission	Launched	Special features
Rossi XTE (NASA)	1995	High time resolution
Chandra (NASA)	1999	High resolution
XMM-Newton (ESA)	1999	High sensitivity
HETE-2 (NASA)	2000	High time resolution